The Chemistry of Fire and Hazardous Materials

THE CHEMISTRY
OF FIRE
AND HAZARDOUS
MATERIALS

Charles F. Turner
President
Hamden Fire Safety Consultants

Joseph W. McCreery
Deputy Chief and Director
Philadelphia Fire Training College

Allyn and Bacon, Inc.
Boston / London / Sydney / Toronto

To all Fire Safety Personnel who have given their lives in the service of others with the prayer that this book will provide the knowledge to assist those who live on to perform their dedicated tasks with greater safety.

art editor: Leora Haywood
series editor: Jeremy Soldevilla
production editor: Margaret Pinette

Library of Congress Cataloging in Publication Data

Turner, Charles F.   1918-
    The chemistry of fire and hazardous materials.

    Includes index.
    1. Combustion.   2. Hazardous substances—
Fires and fire prevention. I. McCreery, Joseph W.,
joint author. II. Title.
QD516.T85   604.7'024628   80-16159
ISBN 0-205-06912-6

Printed in the United States of America.

Contents

Preface

The purpose of this book is not to make chemists out of fire safety personnel nor to make fire safety personnel out of chemists. It was written for fire safety personnel in an attempt to give them a more thorough understanding of the characteristics of the hazardous materials they encounter, and hence the real dangers they so often face. By so doing, it is our hope that we can assist them in doing their jobs with greater safety to themselves and the people they serve.

To this end we have attempted to instruct the interested reader in the chemical background of materials which might be faced in hazardous situations. This requires some knowledge of chemistry and chemical terminology. We have covered only the barest essentials of chemistry in as simple and straightforward a manner as possible, assuming no prior knowledge of chemistry. Readers who apply themselves will gain the necessary knowledge to better protect themselves and others.

No one book can detail the almost infinite number of chemical compounds that exist. We have divided the relatively more common compounds into families and groups so that a reader who comes in contact with a new or unfamiliar compound in the course of work should be able to categorize it and thereby gain at least a general appreciation of the hazards associated with that material.

In order that there be no misunderstanding, we will, at this time, define the terms in our title: *fire* and *hazardous materials*. By "fire" we mean a chemical reaction that produces heat and radiation in some form. The term "hazardous materials" is a bit more difficult to define. Webster defines "hazard" as a source of danger and then the word "danger" as an "exposure to evil, injury, or loss." For our purposes we will leave the "exposure to evil" to the philosophers and theologians and restrict ourselves to the concept of "injury or loss." We will, however, maintain the position that for an event to be hazardous there must be: 1) some form of life or property that is exposed to injury or loss; and 2) an event or events that provide the danger to the safety or well being of that life or property. The event or events may take several forms. There may be direct danger of immediate harm to life or property from physical damage, or there may be indirect danger which can produce delayed harm by means of chemical reactions within the body or property. "Hazardous Materials" therefore would be those materials which would provide a source of danger either directly or indirectly.

We shall emphasize, but not restrict ourselves to, those materials which under fire conditions or the threat of fire present a danger to life or property and would, therefore, be considered hazardous. Chlorine gas, fire or not, presents a tremendous hazard to life and property. Fire safety personnel have been called upon to deal with this gas in leaking tank car situations because no other service is even remotely equipped to deal with it.

A chapter on electricity has been included because in most fire situations electricity is present, and therefore can make many normally nonhazardous metals into very hazardous materials—at least to the touch. Far too many fire personnel have died because, inadvertently, they came in contact with a "hot" conductor.

As the United States increases the pace of its transition toward the metric system, more and more of our lives, both at work and at home, will begin to take on new dimensions—or least unfamiliar dimensions. In an effort to introduce the reader to the metric system, the authors have included metric terms for all important data. The metric number and term are made secondary by placing them in parenthesis. However, the reader must understand that, for the most part, conversions have been made from English measurements to the metric and then rounded mathematically, so as to present practical, whole numbers rather than exact, or nearly exact, conversions carried to several decimal places. The results are practical metric numbers for practical readers.

Finally, because "no man is an island" nor do coauthors alone produce a book, we wish to acknowledge a debt of gratitude to the numerous people whose encouragement and aid have led to this successful conclusion. The authors are indebted to Philadelphia Fire Commissioners Joseph Rizzo and Harry Kite for their continued interest and encouragement; to the men of the Philadelphia Fire Training College and to Battalion Chief Harry Cusick for their suggestions and willingness to play the part of "test students"; to G. Donald Steele, Fire Coordinator for the State of Connecticut, for his many suggestions and continual interest. Special thanks also go to the reviewers, F. Dale Cozad of Eastern Kentucky University, Eugene Fottrell of Atlantic Community College, and Eugene Mahoney of Rio Hondo College for their comments and input during the manuscript's development. The greatest debt, however, is owed to a former teaching colleague, Robert Buggeln. His infinite care in proofreading the manuscript, his excellent suggestions for improvement, and his sincere and continual interest have all made this book a far better one than it would have been without him.

The Chemistry of Fire and Hazardous Materials

Foundations of Chemistry/ 1

HISTORY

Fire has been on earth longer than man. In fact, some theories about the formation of the universe suggest it all may have started with a ball of fire or similar outburst of energy. At the dawn of science, fire was recognized as both friend and foe and was considered to be one of the basic substances that make up the world. Early scientists proposed that there were between two and five basic substances making up everything man could see or feel. The four substances theory was the most popular. According to this theory, everything in the universe consisted of some combination of earth, air, fire and water.

Eventually it became apparent that these things were not basic substances, but were themselves made up of other substances. The first step toward the abandonment of the four substance theory came when the scientists of the twelfth century, called **alchemists**, sought a valid explanation for the behavior of metals. Gold, silver, and copper were known to the alchemists because these metals were found in an almost pure state in nature. Because lead was also well known and far more plentiful, the alchemist claimed to possess secret techniques for the transmutation of lead, the common metal, into the rare metals of royalty: gold, silver, and, in some civilizations, copper. Not only because of their claims, but also because of their knowledge of simple chemical reactions, the alchemists were thought to be magicians.

To the early alchemist, all metals were one substance, differing only in degrees of purity. Therefore, changing one metal into another merely required a purification of the starting metal. The alchemist's failure to transform materials did indicate possible paths to the truth. It was not until the eighteenth century that scientists began a tremendous advance in knowledge by asking: Why do some things turn into fire while others do not? Like anyone seeking to explain a mystery, they offered an explanation which seemed to fit what they were seeing, suggesting that anything that would burn contained a substance called **phlogiston,** a Greek word meaning "the material that burns." The ash, they reasoned, that was left behind after all the phlogiston had escaped in the burning process was merely a residue of the containers that kept the phlogiston locked up until it turned into fire.

Like all scientific theories, the phlogiston theory was subjected to a series of tests of its validity. How could the theorists explain lamp oil, which left no residue? They decided that it must be pure, uncontained phlogiston. Then why did it not escape and turn into fire? Theorists had no answer to that question; the phlogiston theory couldn't stand up to the tests, and was proved unsound.

Like any unsound theory, the phlogiston "answer" satisfied some of the questions, but not *all* of them. To be accepted as scientific, an answer to "why" or "how" must satisfy all; it must be the *only* answer.

An early form of the modern theory about the construction of matter, the atomic theory, was proposed by an Englishman named Robert Boyle in the mid-1600s. It was his belief, after many experiments and observations, that earth, air, fire, and water were combinations of more than one substance, and that the true elements were indivisible substances. He also predicted that many other elements, besides those in earth, air, fire, and water, would someday be discovered. This is a classic example of deductive reasoning.

In Boyle's day, gases were thought to be one continuous substance. However, not long after Boyle proposed his elemental theory, John Dalton explained that several gases can occupy the same space because they are composed of small, "uncut, indivisible" particles which he called **atoms**. Later the Italian physicist Avogadro proposed that these atoms could unite to form particles called **molecules** (meaning "little masses"). Through deductive reasoning, Avogadro suggested that equal volumes of all gases, measured under the same conditions of pressure and temperature, contain an identical number of molecules. Later experiments have proven him correct.

The modern theory of the construction of matter has evolved over the centuries. It has been explained in more and more detail; it has been tested and tested and tested, modified and retested without uncovering a basic flaw in its explanations. Still, it is only a theory. No one has ever seen an atom or a basic molecule. There is still a level of uncertainty. The truth, in this case, cannot be examined directly; it cannot be seen. It must be deduced.

THE SCIENTIFIC METHOD: DEDUCTIVE REASONING

In scientific reasoning, there are distinct differences between levels of certainty.

THE HIGHEST LEVEL: A LAW

A scientific **law** is the explanation of a fact that has been shown to be immutable and unchanging for the area defined. The law of gravity is a familiar example. We know with absolute certainty that gravity is always going to be operable—our whole civilization is built upon that assurance. Our understanding of the mechanics of the law of gravity, how it works, led us to predict that it would *not* be operable in space, away from the earth; that when the first astronauts landed on the moon, the force of gravity there would be only one sixth that of the force of gravity on earth. This is a scientific law—always operating, immutable, a mathematical certainty in its defined area. It is the *only* answer.

THE MIDDLE LEVEL: A THEORY

A scientific **theory** is a proposed explanation for a large group or related series of observations. To be a theory the proposed explanation must have withstood the test of time and experiments without ever having been shown to be false. There must be continued experiment, observation, and complementary explanations answering all pertinent questions in a harmonious fashion without denying any basic, previous observations. It must *seem* to be the only answer.

THE LOWEST LEVEL: A HYPOTHESIS

While a **hypothesis** is barely a level of certainty, it does have some assurance connected with it. It is a reasonable explanation of the results of a series of experiments, attacking the problem from several different angles. Leading up to this level are two steps a scientist must take: **observation** and **experiment.** The entire progression may start with either one or the other, but both must be repeated many times before even the lowest level can be said to have been attained. Furthermore, each experiment and observation must build a single structured answer just as brick on brick builds a great edifice. Observation and experiment must test and complement each other so the hypothesis can be said to be *believed* to be the only answer to the problem.

The steps of scientific reasoning leading through the various levels of certainty are:

1. Ask *why* or *how.*
2. Question the answer given.
3. Propose reasons *why* or *how,* based on *all* the facts.
4. Prove the proposed reasons by experiments.
5. Repeat the steps again and again until the answer can no longer be questioned.

Perhaps a story will help illustrate the scientific method. If in truth Newton was hit upon the head by an apple, his noticing that fact could be called his observation. If Newton shook the tree to see if other apples would hit him or at least fall to the ground, that act could be called one of his experiments. Let us further suppose that when he shook the tree, the only apple that fell off hit him on the head again. His first attempt at an explanation might well have been that apples only fell out of trees when he was standing under them. Being a true scientist, he would, therefore, test his postulation by asking a friend to stand under the tree while he, Newton, shook a branch. When his friend was hit, Newton might have reasoned that apples fall when people stand under apple trees. This, obviously, would not have been the only answer. If he had not noticed it before, by this time Newton would have realized things fall whenever they are not supported. Further reasoning about the bumps on his head, and his friend's head, might have led him to a conclusion, an attempt to explain what he had observed: all objects fall and have weight for the same reason—some sort of attraction toward the earth. After considerably more testing, questioning of conclusions, retesting and thinking about why things are attracted to the earth, the theory of gravity might have evolved. Finally, after considerable time had elapsed with more independent thinking, testing and retesting, always with complementary results and no contradictions, the law of gravity could have been proposed: an attracting force exists which is caused by the earth's mass acting on matter at or near the earth's surface.

Another natural law states that all nature seeks balance. Find some examples of this for yourself.

MEASUREMENT

As a physical science, chemistry is concerned with the structure of matter and the transformations which material substances undergo. Fire chemistry is concerned with the understanding of fire and the reaction of materials when subject to fire. Measurement of the results and changes allows us to specify and quantify our observations.

While there are a great many systems of measurement, the universal language of scientific measurement is the metric system. The metric (from the Greek word metron, a measure) system originated in France in about 1790 and became compulsory there in 1840.

In 1875 the metric system was recognized in an international treaty signed by seventeen countries, including the United States. By 1900 all the major nations of the world and many small ones had officially adopted the metric system, but it was not until 1975 that the Metric Conversion Act (Public Law 94-168) was passed by Congress recommending that the United States join the modern world in its methods of measurement. This act proposed a changeover to the metric system within ten years.

Unlike the cumbersome twelve-inch foot, thirty-six-inch yard and 5280-foot mile of the old English system, the metric measurements are based on the decimal system. Conversions within the system are merely a matter of moving a decimal point to the left to divide and to the right to multiply. The metric system uses a series of prefixes indicating the decimal multiplier.

The more common ones are:

Name	Abbreviation	Multiplier
mega-	M	1,000,000
kilo-	k	1,000
deci-	d	0.1
centi-	c	0.01
milli-	m	0.001
micro-	μ	0.000001
nano-	n	0.000 000 001

A centimeter is therefore one-hundredth part of a meter and a millimeter is one thousandth part of a meter. A milliliter is one thousandth part of a liter. A millisecond is one thousandth part of a second.

TEMPERATURE

There are three common systems for measuring temperature. A fourth scale is used more for scientific calculations (see Figure 1.1). The two most common ones, the Fahrenheit and Celsius scales, are both based on the freezing and boiling temperatures of water. The third scale, called the Kelvin, or absolute, is based on calculations of the hypothetical point at which all molecular activity would cease, and at which point there would be no energy (heat) developed. This point is called **absolute zero**.

A fourth scale, the Rankine, was devised to eliminate the negative numbers of the Fahrenheit scale and to allow the latter to be used for scientific calculations. For this reason it is often referred to as the absolute Fahrenheit scale. Conversion to the Rankine scale is relatively simple:

$$\text{degrees Rankine} = \text{degrees Fahrenheit} + 460$$

To convert from Fahrenheit to Kelvin:

$$\text{degrees Kelvin} = \frac{5}{9}(\text{degrees Fahrenheit} - 32) + 273$$

To convert from Celsius to Kelvin is a bit easier:

$$\text{degrees Kelvin} = \text{degrees Celsius} + 273$$

The Kelvin scale is to the Celsius scale what the Rankine scale is to the Fahrenheit scale.

The metric system uses the Celsius scale for temperature. This scale is also known as the centigrade scale because there are one hundred divisions between the freezing point and boiling point of water.

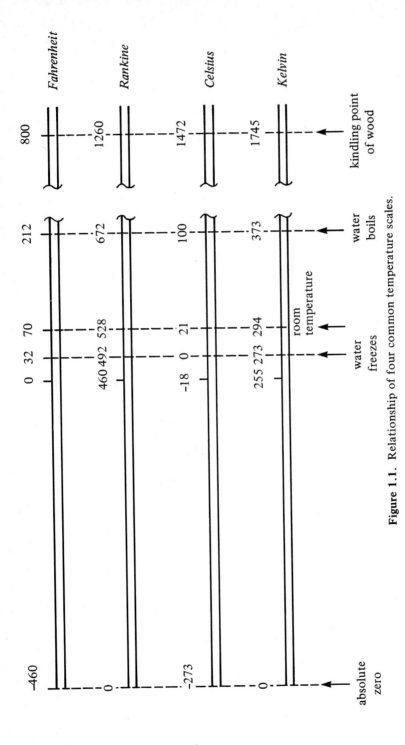

Figure 1.1. Relationship of four common temperature scales.

Conversions between the Fahrenheit and Celsius scales can be accomplished by using the following formulas.
If the Fahrenheit temperature is known:

$$\text{degrees Celsius} = (\text{degrees Fahrenheit} - 32) \, \frac{5}{9}$$

example: What Celsius temperature is equal to 70 °F?

$$°C = (°F - 32) \, \frac{5}{9}$$

$$= (70 - 32) \, \frac{5}{9}$$

$$= (38) \, \frac{5}{9} = \frac{38 \times 5}{9} = \frac{190}{9}$$

$$°C = 21 \text{ (answer)}$$

If the Celsius temperature is known:

$$\text{degrees Fahrenheit} = \left(\text{degrees Celsius} \times \frac{9}{5}\right) + 32$$

example: What Fahrenheit temperature is equivalent to the boiling point of ethyl alcohol, which is 78.5 °C?

$$°F = \left(°C \times \frac{9}{5}\right) + 32$$

$$= \left(78.5 \times \frac{9}{5}\right) + 32$$

$$= \frac{78.5 \times 9}{5} + 32 = \frac{706.5}{5} + 32$$

$$°F = 173.3 \text{ (answer)}$$

Most people will not be able to do the above calculations without a calculator or at least a pencil and paper, but for quick approximations the following mental calculations can be used. In the conversion from Fahrenheit to Celsius the fraction that causes the trouble is the $5/9$. Note that this fraction is however very close to $5/10$ or $1/2$. When converting from degrees Fahrenheit to degrees Celsius in the above example we mentally subtract 32 from 70 to get 38. One half of 38 is 19, which is an approximation of the correct answer. For those who wish to be more exact 10 percent may be added to that answer. Thus 10 percent of 19 is 1.9 and $19 + 1.9 = 20.9$. This is rather close to 21.

In converting from Celsius to Fahrenheit the difficulty-causing fraction is $9/5$. This is close to $10/5$ or 2. Therefore in this case multiply the Fahrenheit temperature by 2 ($79 \times 2 = 158$). Adding 32 to this results in 190. Again, this is about 10 percent in error. We then subtract 10 percent of 190, or 19, to get 171. This is rather close to 173.

VOLUME

The metric system greatly simplifies the problem of volume relationships and the conversions from one to the other. Instead of pints, quarts, and gallons, all with different relationships, the metric system uses one unit of volume: the **liter**. Smaller units of the liter are merely decimal parts of the liter so that conversions can be made by merely moving the decimal point.

For example,

$$1 \text{ liter} = 1000$$
$$1 \text{ milliliter} = 0.001 \text{ liters}$$
$$10 \text{ milliliters} = 0.01 \text{ liters}$$
$$100 \text{ milliliters} = 0.1 \text{ liters}$$

Conversion to our system is a bit more difficult. A liter is about equal to 1.06 quarts; 1 quart is equal to about 0.946 liters. Common conversion factors[1] between the two systems are:

To Convert From	To	Multiply By	To Convert From	To	Multiply By
quarts	liters	0.95	liters	quarts	1.06
gallons	liters	3.8	liters	gallons	0.26
gallons/minute	liters/minute	3.8	liters/minute	gallons/minute	0.26
cubic feet	cubic meters	0.03	cubic meters	cubic feet	35.0

WEIGHT AND MASS

In the nonscientific community, the terms weight and mass are often used interchangeably, but to the scientist they are quite different. **Weight** properly refers to the downward force caused by a body and is due to the *force* of gravity acting upon it. **Mass** has nothing to do with gravity, referring only to the *amount* of matter in a body. The weight of a body would be different on earth than on the moon because the force of gravity on the moon is only about one-sixth that on the earth. The mass of body would be the same on the earth as on the moon.

A hose coupling that has a mass of six pounds also has a weight of six pounds at sea-level on earth. On the surface of the moon, while its mass would still be six pounds, the weight of the coupling would be only one pound because the force of gravity on the moon is about one-sixth that of the earth. On a different planet the mass of that coupling would still be six pounds but the weight would depend on the force of gravity on that planet.

---

[1]In all conversion tables in this chapter, most conversions will be close approximations rather than scientifically exact.

The metric system derives its unit of weight, the **gram**, on the basis of volume. In the metric system the weight of 1 liter of pure water at 4 °C is equal to 1000 grams or 1 kilogram.

When converting between our present system and the metric, or reverse, the following factors should be used:

To Convert From	To	Multiply By	To Convert From	To	Multiply By
ounces	grams	28	grams	ounces	0.035
pounds	kilograms	0.45	kilograms	pounds	2.2
tons (2000 lb)	kilograms	907.2	kilograms	tons (2000 lb)	0.001
tons (2000 lb)	tonnes	0.9	tonnes	tons (2000 lb)	1.1

LENGTH

In the United States, the average person must cope with distances measured in inches, feet, yards, and miles, the scientist measures distances in **meters**. Originally the meter was defined as one ten-millionth part of the distance from the equator to the North Pole on a line through Dunkirk, France, and Barcelona, Spain. The standard meter is now a solid bar of platinum kept at the Bureau of Metrology near Paris. Platinum was chosen because it expands or contracts the least of all metals with changes in temperature.

Long distances are recorded in kilometers, small distances in centimeters (100 to a meter) and very small distances in millimeters, abbreviated mm (1000 to a meter).

Conversions can be made with the following factors:

To Convert From	To	Multiply By	To Convert From	To	Multiply By
inches	millimeters	25.4	millimeters	inches	0.04
inches	centimeters	2.54	centimeters	inches	0.4
feet	centimeters	30	centimeters	feet	0.033
feet	meters	0.3	meters	feet	3.3
yards	meters	0.9	meters	yards	1.1
miles	kilometers	1.61	kilometers	miles	0.62

Diameter is a form of length. Some common hose diameter conversions are shown in Figure 1.2.

AREA

Area is a length multiplied by a length and is therefore a squared term in both the English and metric systems.

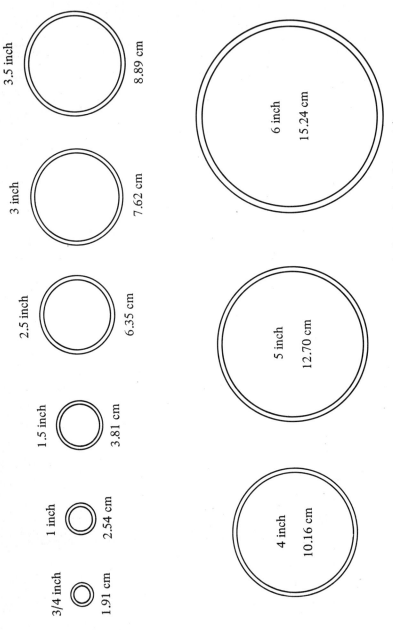

Figure 1.2. Common hose diameter equivalents: standard to metric.

To Convert From	To	Multiply By	To Convert From	To	Multiply By
square inches	square centimeters	6.5	square centimeters	square inches	0.16
square feet	square meters	0.09	square meters	square feet	10.8
square yards	square meters	0.8	square meters	square yards	1.2
square miles	square kilometers	2.6	square kilometers	square miles	0.4

PRESSURE

Pressure is defined as a force applied over an area. An example of this is air pressure. It has been said that we are living at the bottom of a great sea of air. The weight of that sea of air exerts a force of about 14.7 pounds on every square inch of earth. That is a force applied over an area—14.7 pounds per square inch (psi).

In the metric system there have been several different terms for pressure, including millimeters of mercury, kilograms per square centimeter, and bars. These have all been dropped in favor of the term pascal (pronounced like "rascal"), named for a French scientist. The pascal is a very small unit: 1 psi equals 6897 pascals or 6.897 kPa.

Because other units are still in use and will be for some time the following conversions are given:

To Convert From	To	Multiply By	To Convert From	To	Multiply By
psi	bars	0.069	bars	psi	14.5
psi	kg/cm²	0.070	kg/cm²	psi	14.2
psi	mm of mercury	51.71	mm of mercury	psi	0.02
psi	pascals	6897	pascals	psi	0.00015

SI NOTATION

In 1960 a General Conference on Weights and Measures modernized and categorized the metric system by agreeing upon an International System of Units. In French the first two words are "Systeme Internationale," which is abbreviated as SI.

The basic units in the SI are:

Measurement	Name	Symbol
length	meter	m
weight	kilogram	kg
time	second	s
electric current	ampere	A
temperature	Kelvin	K

Although scientists have adopted the Kelvin scale there is almost universal use of the Celsius scale for nonscientific work.

From the basic SI units many others can be derived. Some of the more common ones are:

Measurement	Name	Symbol	In Terms of Basic Units
frequency	Hertz	Hz	units/sec
force	Newton	N	$kg\ m/s^2$
pressure	Pascal	Pa	$kg\ m/s^2/m^2$
energy	Joule	J	$(kg\ m/s^2)m$
power	Watt	W	$(kg\ m/s^2)m/s$

POWERS OF TEN

Because scientists often use very large or very small numbers, a mathematical tool is used to reduce those numbers to a more easily manipulated group of digits. For example, chemists have determined that the number of molecules in a standard volume (22.4 liters) of gas at standard temperature (0 °C) and pressure (760 mm of mercury) is 602,000,000,000,000,000,000,000 or 602 sextillion. The chemist would write this 6.02×10^{23}.

This is a decimal system based on the number ten multiplied by itself some number of times. This number of times is written as a right hand superscript, called an **exponent,** and the ten is then said to be raised to that power. The principal digits of the number to be reduced are then written as multiplied by ten to an appropriate power.

Ten raised to a power is:

$$10^0 = 1$$
$$10^1 = 10$$
$$10^2 = 100$$
$$10^3 = 1000$$
$$10^4 = 10,000$$
$$10^5 = 100,000$$

Note that with the positive powers, such as we have here, the number of zeros after the 1 each time is equal to the exponent. Using the exponential system, the number 2000 could be written:

$$2 \times 10^3 = 2 \times 1000 = 2000$$

or

$$20 \times 10^2 = 20 \times 100 = 2000$$

The number 51,600 could be written:

$$5.16 \times 10^4 = 5.16 \times 10,000 = 51,600$$

or $$51.6 \times 10^3 = 51.6 \times 1000 = 51,600$$

or $$516 \times 10^2 = 516 \times 100 = 51,600$$

In dealing with numbers less than one, negative powers are used in much the same way to enlarge the number for practical handling.

$$10^0 = 1$$
$$10^{-1} = 0.1$$
$$10^{-2} = 0.01$$
$$10^{-3} = 0.001$$
$$10^{-4} = 0.0001$$
$$10^{-5} = 0.00001$$
$$10^{-6} = 0.000001$$

Note that with negative powers the number of zeros is one less than the exponent and the zeros are to the left of the one.

Using exponential notation, 0.0001 could be written:

$$1 \times 10^{-4} = 1 \times 0.0001 = 0.0001$$

The number 0.000000233 could be written:

$$2.33 \times 10^{-7} = 2.33 \times 0.0000001 = 0.000000233$$

or $$23.3 \times 10^{-8} = 23.3 \times 0.00000001 = 0.000000233$$

The conventional way of writing a number is to place the decimal point so that there are one or two digits to the left of the point and then using an exponent of the proper power, either positive or negative.

Numbers using the exponential system may be multiplied and divided quite easily. To multiply, the principal numbers are multiplied in the usual manner and the exponents are added:

example:

$$(2.02 \times 10^{17})(3.1 \times 10^2) =$$

Multiply the principal numbers:

$$2.02 \times 3.1 = 6.26$$

Then add the exponents:

$$17 + 2 = 19$$

Therefore $$(2.02 \times 10^{17})(3.1 \times 10^2) = 6.26 \times 10^{19}$$

To divide numbers with exponents, divide the principal numbers in the normal manner and subtract the exponents.

example:
$$\frac{6.26 \times 10^{19}}{3.1 \times 10^2} =$$

Divide the principal numbers:

$$6.26 \div 3.1 = 2.02$$

Subtract the exponents:

$$19 - 2 = 17$$

Therefore:
$$\frac{6.26 \times 10^{19}}{3.1 \times 10^2} = 2.02 \times 10^{17}$$

example:
$$\frac{5.43 \times 10^4}{2.3 \times 10^{12}} =$$

$$5.43 \div 2.3 = 2.36$$

$$4 - 12 = -8$$

Therefore:
$$\frac{5.43 \times 10^4}{2.3 \times 10^{12}} = 2.36 \times 10^{-8}$$

When the mathematical function required is addition or subtraction itself, the powers to which each ten is raised must be made equal before the principal numbers are added. The exponents then remain the same.

example: $\qquad (48 \times 10^4) - (32 \times 10^2) = ?$

Convert $\qquad 48 \times 10^4$ to 4800×10^2

Then $\qquad (4800 \times 10^2) - (32 \times 10^2) = 4768 \times 10^2$, or 47.68×10^4

example: $\qquad (37 \times 10^3) + (24 \times 10^2) = ?$

Convert $\qquad 37 \times 10^3$ to 370×10^2

Then $\qquad 370 + 24 = 394 \times 10^2$, or 39.4×10^3

Converting exponents back to normal numbers is not difficult if two simple rules are followed:

1. When the exponent is positive move the decimal point to the right a number of places equal to the power.
2. When the exponent is negative move the decimal point to the left a number of places equal to the power.

REVIEW QUESTIONS

1. Detail the phlogiston theory and the reasons for its failure.
2. Who is credited with the origin of the modern atomic theory? How did he progress to its hypothesis?
3. What were John Dalton's contributions to our modern atomic theory?
4. State Avogadro's law concerning quantities of molecules in gases.
5. Detail and explain the steps in deductive reasoning.
6. Write a paragraph on the origins of the metric system.
7. Give the abbreviation and multiplier for the following metric prefixes: mega-, kilo-, centi-, milli-, deci-, micro-, nano-.
8. Convert 0 $^\circ$C, 20 $^\circ$C, 100 $^\circ$C, 456 $^\circ$C to Fahrenheit and Kelvin scales.
9. Convert the following line diameters into metric measurement: ¾ inch, 1 inch, 1½ inches, 2½ inches, 3½ inches.
10. Multiply:
    a. $10^2 \times 10^3$
    b. $10^{-2} \times 10^2$
    c. $(2.02 \times 10^{12}) \times (1.7 \times 10^{10})$
11. Divide:
    a. 10^5 by 10^3
    b. 10^{-2} by 10^4
    c. 7.8×10^4 by 2.9×10^2
12. Discuss how the metric system affects firefighters. Consider length of ladders, etc.

Basic Inorganic Chemistry/ 2

Fire chemistry is, like any other chemistry, the study of the interrelationships of molecules and atoms. All fields of chemistry follow the same natural laws; fire chemistry is a study of reactions and results in one specific area.

MODERN ATOMIC THEORY

The modern theory about the construction of the atom suggests that all matter is composed of atoms and that all atoms are made up of positively and negatively charged masses in very precise structures with an arrangement that is unique for each element. The negatively charged masses are called **electrons**; the positively charged masses are called **protons**. Experiments probing this theory have led to the discovery of other masses within the atom, but for our purposes, these masses are not important.

The precise arrangement proposed for the atom is most simply described as a core or nucleus made up of one or more protons with one or more electrons orbiting around the core like planets around the sun.

The theory proposes that the electrons have very little weight and that most of the weight of the atom is in the core. How do we explain how atoms with many of the same physical and chemical characteristics still have different weights? It was suggested that there are other particles or masses within the atom which have weights similar to the proton. However, while these masses have weight, they can have no charge, either positive or negative, without upsetting the electrical balance of the atom. Exhaustive tests have shown that the atom is balanced, as is all nature, and that the neutral masses, called neutrons, do exist.

According to the theory, the number of protons in the core is always a whole number and distinct for each element. For this reason the number of protons is recorded as the **atomic number.**

It has been further shown that the electrons must be in orbits or shells, and that these shells contain very definite numbers of electrons. The first shell, nearest to the core, contains a maximum of two electrons, the next eight, then eight, eighteen, thirty-two, fifty, and so on. Experiments indicate that in the lighter atoms, each shell will fill up to its maximum before the next shell begins to fill. With the heavier atoms, this does not hold true.

These shells are of increasing diameter with the energy levels of these shells increasing as they get farther from the nucleus. Experiments indicate there are subshells with varying electron rotations and spins.

ORDER IN THE ELEMENTS

In the mid-1800s, two scientists working independently in Russia and Germany noticed that the physical and chemical characteristics of the then-known elements seemed to be related. For example, one group reacted with almost all other elements, while another group exhibited no tendency to react with any other elements. It was also noticed that those groups could be arranged in a table into families because of other physical properties. There is, for example, a family of gases that will not usually react with other elements. These gases are said to be inert; they include helium, neon, argon, krypton, xenon, and radon. They have similar, but not identical, physical and chemical properties and can be arranged in a column roughly according to weight. Another group exhibiting family resemblances can also be arranged in a column according to weight but have quite different physical and chemical characteristics than the first group. These are very reactive, and include fluorine, chlorine, bromine, and iodine. There are a number of groups or families of elements with similar properties. The recurrence of similarities at regular intervals of periods in the sequence of the elements is called the **periodic law**.

OUTER SHELL ELECTRONS

Modern atomic theory proposes that elements will unite when an atom, either alone or as part of a molecule, comes into immediate contact with one or more atoms under the proper conditions, and the two atoms bond together. Sometimes there is an outer orbital electron transfer, and sometimes there is not. In cases where the electron is transferred, one atom loses the electron or electrons while the other atom gains the lost electron or electrons. The potential for either exchanging or sharing electrons is called **valence**, while the electrons that can be shared, gained, or lost are called **valence electrons**. There are two basic types of valence: **electrovalence**, which involves gain or loss; and **covalence**, which is the sharing of the electron mass. The situation can be compared to a kid who brings the baseball into a pick-up game; it is still his ball but everyone else is using it to play the game, and he is part of that game. In our case, the game would be the new material formed by the union of the atoms or molecules, and the ball would be the electrons that are holding it together. The sharing of the valence electrons

METALS — NONMETALS

PERIODS	IA	IIA	IIIB	IVB	VB	VIB	VIIB	VIII			IB	IIB	IIIA	IVA	VA	VIA	VIIA	O
1	1.0079 **H** 1																	4.00260 **He** 2
2	6.94 **Li** 3	9.01218 **Be** 4											10.81 **B** 5	12.011 **C** 6	14.0067 **N** 7	15.9994 **O** 8	18.9984 **F** 9	20.179 **Ne** 10
3	22.9898 **Na** 11	24.305 **Mg** 12											26.9815 **Al** 13	28.086 **Si** 14	30.9738 **P** 15	32.06 **S** 16	35.453 **Cl** 17	39.948 **Ar** 18
4	39.098 **K** 19	40.08 **Ca** 20	44.9559 **Sc** 21	47.90 **Ti** 22	50.9414 **V** 23	51.996 **Cr** 24	54.9380 **Mn** 25	55.847 **Fe** 26	58.9332 **Co** 27	58.71 **Ni** 28	63.546 **Cu** 29	65.38 **Zn** 30	69.72 **Ga** 31	72.59 **Ge** 32	74.9216 **As** 33	78.96 **Se** 34	79.904 **Br** 35	83.80 **Kr** 36
5	85.4678 **Rb** 37	87.62 **Sr** 38	88.9059 **Y** 39	91.22 **Zr** 40	92.9064 **Nb** 41	95.94 **Mo** 42	98.9062 **Te** 43	101.07 **Ru** 44	102.9055 **Rh** 45	106.4 **Pd** 46	107.868 **Ag** 47	112.40 **Cd** 48	114.82 **In** 49	118.69 **Sn** 50	121.75 **Sb** 51	127.60 **Te** 52	126.9046 **I** 53	131.30 **Xe** 54
6	132.9054 **Cs** 55	137.34 **Ba** 56	57–71 *	178.49 **Hf** 72	180.9479 **Ta** 73	183.85 **W** 74	186.2 **Re** 75	190.2 **Os** 76	192.22 **Ir** 77	195.09 **Pt** 78	196.9665 **Au** 79	200.59 **Hg** 80	204.37 **Tl** 81	207.2 **Pb** 82	208.9804 **Bi** 83	(210) **Po** 84	(210) **At** 85	(222) **Rn** 86
7	(223) **Fr** 87	(226.0254) **Ra** 88	89–103 †	(260) **Ku** 104	(260) **Ha** 105													

TRANSITION METALS

* LANTHANIDE SERIES

138.9055 **La** 57	140.12 **Ce** 58	140.9077 **Pr** 59	144.24 **Nd** 60	(145) **Pm** 61	150.4 **Sm** 62	151.96 **Eu** 63	157.25 **Gd** 64	158.9254 **Tb** 65	162.50 **Dy** 66	164.9304 **Ho** 67	167.26 **Er** 68	168.9342 **Tm** 69	173.04 **Yb** 70	174.97 **Lu** 71

† ACTINIDE SERIES

(227) **Ac** 89	232.0381 **Th** 90	231.0359 **Pa** 91	238.029 **U** 92	237.0482 **Np** 93	(242) **Pu** 94	(243) **Am** 95	(245) **Cm** 96	(245) **Bk** 97	(248) **Cf** 98	(253) **Es** 99	(254) **Fm** 100	(256) **Md** 101	(253) **No** 102	(257) **Lr** 103

Figure 2.1 Periodic Table of the Elements.

is done in a number of different ways and the **bond** formed is given a title accordingly. The common sharing type bonds are the **covalent bond** and the **coordinate covalent bond.** Note that the loss, gain, or sharing of electrons always takes place among the outer orbital electrons. When this bonding occurs, one or more electrons are involved, and a new molecule results. The properties of the new entity are completely different from any of those of the original elements or molecules. The element sodium, for example, is a soft, malleable, ductile substance and is one of the most reactive metals known. The element chlorine is a moderately reactive greenish-yellow gas that is toxic to humans. Despite these properties, the molecule that is formed from the chemical reaction of sodium and chlorine is one of the most common and useful substances on earth. It is, of course, sodium chloride, which, in its refined form, we call table salt.

INORGANIC CHEMICAL REACTIONS

The union of two or more elements is *not,* however, the only way in which elements react with each other. There are four basic types of chemical reactions.

· In **composition reactions,** two or more materials unite to form a single new compound.

If A and B were elements, then this type of reaction could be expressed:

$$A + B \longrightarrow AB$$

example: carbon + oxygen ⟶ carbon dioxide

The products of composition type reactions, molecules, have another name: **compounds.** Compounds may react chemically with other compounds to form yet other compounds, often of vastly different properties.

· A **decomposition reaction** is just the reverse of a composition reaction and could be expressed:

$$AB \longrightarrow A + B$$

example: ammonium hydroxide ⟶ ammonia + water

· In **simple replacement reactions,** one element displaces another in a compound:

$$A + BC \longrightarrow B + AC$$

example: zinc + hydrogen chloride ⟶ hydrogen + zinc chloride

· In **double replacement reactions,** two compounds react to form two new compounds by the exchange of partners:

$$AB + CD \longrightarrow AD + CB$$

example: lead sulfide + hydrogen chloride $\longrightarrow$ lead chloride + hydrogen sulfide

The principal actors in the above four types of chemical reactions are the valence electrons. Because the number of outer shell or valence electrons that can be given up or taken up varies with the outer shell configuration, the type of valency of the element or molecule will vary.

A few examples of the various types of valence are:

Element	Valence Type	Electrons Shared	Typical Diagram
Na	monovalent	1	Na — Cl
O	divalent	2	$O\langle^H_H$
N	trivalent	3	$H - N - N\langle^H_H$ with H below first N
C	tetravalent	4	$H - C - H$ with H above and H below

Some elements such as nitrogen will exhibit **polyvalency.** That is, depending on the combining element or elements, these elements will show more than one type of valency. Nitrogen, for example, can share either two, three, or five electrons.

These various types of valency are also indicated in the shorthand which chemists use to describe what takes place during a chemical reaction.

For example, when monovalent hydrogen reacts with divalent oxygen, it requires two atoms of the monovalent element to satisfy one atom of the divalent material.

$$H_2 + O \longrightarrow H_2O$$

Translating this shorthand, the chemist understands that two atoms of hydrogen (monovalent $2 \times 1 = 2$) react with one atom of oxygen (divalent $1 \times 2 = 2$) to form one atom of water.

Copper, as with several other metals, can exhibit polyvalency. Reacting with sulfur, it can be either monovalent or divalent. Sulfur is always divalent; so, in one case:

$$Cu + S \longrightarrow CuS$$

This means that one atom of copper reacts with one atom of sulfur to form one atom of cuprous sulfide. In another case:

$$2Cu + S \longrightarrow Cu_2 S$$

This notation can be translated to mean that two atoms of copper react with one atom of sulfur to form one atom of cuprous sulfide.

Note that when an element has two valency forms, the compound formed from the higher valency atoms is distinguished from the other (lower) by an -ic ending to its name. The compound formed from the lower valence atom is given an -ous ending to its name. With nitrous oxide and nitric oxide, the latter is formed from the higher valence nitrogen.

To remember which is which:

I C

O U S

The pyramid indicates the "ic" is higher (valence) than "ous."

NAMES, SYMBOLS, AND FORMULAS

The origin of the names of early chemical elements is lost in antiquity. Newly discovered elements are usually named after a person or place.

Very early in history of chemistry, the scientists of that day, to save time and scarce writing materials, devised a chemical shorthand, today referred to as **chemical symbols.** Because the universal language of the early scientists was Latin, the chemical shorthand was an abbreviation of the Latin names, as applied to the elements known at that time.

English Translation	Latin Name	Symbol
gold	aurium	Au
silver	argentum	Ag
iron	ferrum	Fe
lead	plumbum	Pb
mercury	hydrargyrum	Hg
sodium	natrium	Na
copper	cuprum	Cu

Elements discovered more recently, such as berkelium, californium, einsteinium, and curium are given endings in the early tradition to produce Latin-like names and symbols to match: Bk, Cf, Es, and Cm, respectively.

STOICHIOMETRY

The proper combination of reactants is very important not only for ease of reaction but for consideration of the amount of energy that will be given off as well as the amount of reaction products that will be produced. In every chemical reaction there is an optimum combination of components that will produce the most efficient reaction. In the specific case of some gas and air reactions, it is common to have either too much fuel (gas) and not enough air or conversely, too much air and not enough gas. The optimum ratio of reactants that utilizes all of the reactants and produces the maximum output of reaction products, including energy in the form of heat and light, is called the **stoichiometric mixture**.

A mixture of marsh gas and air is flammable in any proportion between about 5 percent and 15 percent methane. As that upper limit of 15 percent marsh gas is approached, the reaction will proceed more and more slowly and with a very sooty, yellow flame. In this mixture the combination is said to be fuel rich, the heat output is relatively low, and the production of solids (carbon) is quite high. If the mixture is controllable, and the marsh gas content of the ratio continually increased, the flame will go out. The reaction stops. Changing the ratio in the opposite direction, as the lower limit of 5 percent marsh gas is approached, the reaction is said to be oxidizer rich and tends to snuff out easily.

When ignited at a ratio of 9.5 percent marsh gas and 90.5 percent air, there is complete combustion of the marsh gas in a violent explosion along with a maximum heat output. This is the stoichiometric mixture for this reaction at normal pressure. The stoichiometric mixture is influenced by surrounding pressure.

It should not be inferred that all stoichiometric mixtures are explosive; they are not. The stoichiometric ratio for reacting iron ore (iron oxide) with carbon is as follows:

$$Fe_2O_3 + 3C \longrightarrow 2Fe + 3CO$$

The ratio therefore is one part iron oxide to three parts of carbon. (Chemists do not write the number one in chemical reactions; it is not needed.) The important concept here is that of the ratio. The number three in front of the carbon as well as the unwritten number one might stand for any unit of quantity: molecules, grams, pounds, tons, or anything else *as long as* all other numbers preceding the molecules or atoms have the same units.

While the term stoichiometric ratio is an exact term, it is related to other terms that are broader in scope and of more interest to fire personnel. These are the terms combustion ratio, flammability or ignition limits, and combustion range. Usually the midpoint of the limits of these terms is at the stoichiometric point. They are all related to the chemical equation and the efficiency of the combustion as related to the fuel and oxidizer ratio.

PHYSICAL AND CHEMICAL CHANGE

That materials seem to react when mixed together does not, in itself, indicate that they have chemically reacted. When sodium chloride is placed in water, it will apparently react and disappear. In truth, however, there has been no chemical change. The properties of both the water and the salt are substantially the same, and both of the components, salt and water, can be recovered by fairly simple means: evaporation of the water will leave the salt while the water can be recovered by condensing the steam during the evaporation process. The term chemists use for the nonreactive combining of compounds or elements is **mixture**. The key difference between a mixture and the formation of a new compound by chemical reaction is the change in properties which occurs because of the chemical reaction.

The four sign posts of chemical reaction are changes in composition, chemical energy, physical properties, and chemical properties.

Examples of physical changes as opposed to chemical changes are:

freezing of water	physical
charging a storage battery	chemical
dissolving salt in water	physical
reacting sodium and chloride	chemical
sawing wood	physical
burning wood	chemical

In a practical sense, the firefighter facing hazardous fire situations is usually dealing with chemical reactions, which involve the gain or loss of electrons. In these reactions, one atom will give and one will receive even though both may be using the electron in the new compound. Somewhat like the boy who brought the ball into that baseball game and was then put out in the field; he is part of the game, but he doesn't get very close to his ball very often. The reactions are called oxidation-reduction or **redox** reactions. The gain of electrons is properly called **reduction** and the loss is called **oxidation**. The exact reasons for this go deep into reaction chemistry and will not be dealt with here. Nevertheless, we shall be discussing oxidation in many and varied forms in later chapters, so it will be well to remember which is oxidation and which is reduction. One way is to remember that oxidation starts with an "O" and the word loss has an "O" in it thus:

L
O X I D A T I O N
S
S

ENERGY

Energy exists in two forms, **kinetic** and **potential**. Kinetic energy is the energy of motion. A fire truck racing to a fire at 60 mph is a good example of kinetic energy. The measure of kinetic energy is proportional to the mass of the matter times its velocity squared. It can be expressed mathematically:

$$\text{K.E.} = \frac{1}{2} MV^2$$

But:
$$\text{Mass} = \frac{\text{Weight}}{\text{Acceleration due to Gravity (32.15)}}$$

Therefore, substituting for M in the K.E. equation:

$$\text{K.E.} = \frac{1}{2} \frac{W \times V^2}{32.15}$$

or
$$\text{K.E.} = \frac{W \times V^2}{64.30}$$

A 20,000 pound rig roaring down the highway at 60 mph would possess about 2,408,710 foot pounds of kinetic energy. That is the same amount of kinetic energy as 2,408,710 pounds falling one foot or one pound falling 2,408,710 feet. The velocity squared term is scientific proof that speed develops the energy that can kill on the highway.

That same piece of apparatus parked at the top of a hill, its brakes set, would possess potential energy because of its *position* at the top of the grade. If the brakes should suddenly fail, and the rig began to roll down the hill, its potential energy would then be undergoing change to kinetic (motion) energy. A piece of apparatus stopped at the foot of the hill would possess less potential energy than the same apparatus at the top of the hill because its position is lower. Potential energy is often referred to as the "stored" energy. *PO*tential energy is the energy of *PO*sition.

Chemical energy is a type of potential energy, not because of the position of the chemical substance in a bottle on a shelf, *but* because of the position of its orbiting electrons. This potential energy can be released to become kinetic energy derived from a chemical reaction. When a chemical reaction takes place, the materials which react (each possessing its own level of chemical energy) unite to form a new substance with new properties. One of these new properties is the quantity of chemical energy the new substance possesses. Every chemical change results in either an *excess* of energy, which is given off as the reaction proceeds, or in a *lack* of energy, which must be replaced if the reaction is to be kept going. A fire or an explosion is the result of potential chemical energy being transformed into kinetic energy with the escape of the excess energy.

When there is an *ex*cess of energy and, therefore, energy *ex*its from the reaction, the reaction is said to be of the **exothermic** type. When energy must be continuously put *into* the reaction to keep it going, the reaction is said to be of the **endothermic** type.

A burning house is an example of an exothermic chemical reaction in which the wood is uniting with oxygen rapidly enough to produce flame and heat (energy). It would be a mistake to claim that because someone had to start that reaction with a match, that it was, therefore, endothermic. The chemical reaction between the wood and oxygen did not start until the flame and heat were given off. It is the *reaction* that we are characterizing, and wood burning is *exothermic.*

A familiar example of an endothermic reaction is that of recharging a storage battery. In the recharging reaction, the charger supplies electrical energy to change, by means of a chemical reaction, the lead sulfate into lead oxide, lead, and sulfuric acid, in the presence of water. Remove the energy source (charger) and the reaction stops. Because the energy must be continuously supplied to keep the reaction going, the reaction is *endothermic.*

One thing that we must consider now is the difference between heat and temperature. **Temperature** is a measure of the *intensity* of the chemical energy. **Heat** is a measure of the *quantity* of energy. A wooden match being used to light a cigarette will burn at the same temperature as that of a large frame house because it is basically the same chemical reaction:

$$wood + oxygen = heat + flame + gases$$

However, the frame structure will give off far more heat than the single match.

While a **degree** is the term used to describe the unit of temperature, the term used to specify the amount of heat is the **calorie**. One calorie is the amount of heat needed to raise the temperature of one gram of water one degree Celsius from 14.5 °C to 15.5 °C. Strange as it may seem, the amount of heat needed to raise the temperature one degree is not always the same for water temperatures higher or lower than this. The chemical reaction that occurs when a wooden match burns develops about 500 calories. One thousand calories is equal to one kilocalorie, abbreviated Kcal.

STATES OF MATTER

We have been discussing elements, atoms, molecules, and the reactions between them, without consideration of the forms those masses of atoms or molecules appear to have before or after the reaction. The visible forms, which the masses of like atoms or molecules take, are known as **states of matter**. These states of matter are really descriptions of the relationships of the atoms or molecules.

There are three states of matter: **solid, liquid,** and **gas.**

SOLIDS

In the solid state, molecules are packed together in a great crunch resulting in maximum density. The molecules are very dynamic in themselves and have great spaces within their own fields. Because of their relative closeness, there are tremendous forces interacting between the atoms or molecules, so that solids are characterized by the ability to retain a definite shape, to possess rigidity and volume.

LIQUIDS

In liquid state, the molecules are packed together more loosely than in solids. There are greater spaces between them, and the intermolecular forces are not so great. Because of this liquids have less rigidity and do not assume a shape of their own, but adapt themselves to the shape of their enclosure. If there is no enclosure, they tend to spread out over a wider and wider area, restrained only by their own internal forces.

GASES

In gaseous state, the molecules are far apart. There is a lot of space between them, as there is in their internal makeup. The intermolecular forces are much lower than in either solids or liquids, so gases have no rigidity and therefore assume both the shape and volume of their enclosure whether it be a bottle, a tank, or the atmosphere. Gas molecules are very dynamic with constant movement. Gases have other unique properties. They can be compressed when energy is put into them and will always expand to fill their container. When compressed or expanding, the gases exert a force called **pressure** on the sides of the vessel. Pressure is the direct result of the dynamic molecular activity going on within a container. More energy put into the gas by compression or heating makes the molecules become more excited and move faster, and the more they bounce off each other and off the container walls, the higher the pressure becomes. Automobile or truck tires always show an increase in pressure after running on the road because the friction developed with the road surface results in an increase in the heat (energy) in the tire. When this energy gets to the gas (air) inside the tire, it excites the molecules. They increase their velocities; more collisions occur between molecules as well as with the container walls. We read this increased activity as a higher pressure.

      The behavior of gases under variations of temperature, pressure, and volume is described very exactly by a series of mathematical equations known as the **gas laws**. These laws hold for normal pressures and temperatures. At extremely high pressures and temperatures these simple formulas must be modified, but at everyday temperatures and pressures, the changes in volume of a gas with pressure can be expressed by Boyle's law, named for the man who first demonstrated it. Boyle's law states that when the temperature is held constant, the volume of a given amount of gas varies indirectly with the pressure. In other words, an increase in volume results in a decrease in pressure and increase in pressure results in a decrease in volume, so long as the temperature is held constant. Mathematically it is expressed:

$$P_1 \times V_1 = P_2 \times V_2 \qquad (t \text{ constant})$$

In the above formula, P_1 and V_1 are the starting or original pressure and volume, while P_2 and V_2 are the pressure and volume which result after a change in one or the other, as long as there is no change in the temperature of the gas.

example: If the original pressure is one atmosphere on a volume of three cubic feet of gas and it is then compressed very slowly (no temperature change) to two cubic feet, what would the new pressure be?

$$P_1 \times V_1 = P_2 \times V_2$$
$$1 \times 3 = P_2 \times 2$$
$$3 = 2P_2$$

Now solve for P_2:

$$2P_2 = 3$$
$$P_2 = \frac{3}{2}$$
$$P_2 = 1.5 \text{ atmospheres} \quad (\text{answer})$$

When we wish to determine how the volume of a gas changes with absolute temperature, we use Charles's law, named for a French chemist. This law states that, holding the pressure constant, the volume of a gas varies directly with the absolute temperature. In short, as the temperature goes up the volume goes up. Mathematically:

$$\frac{V_1}{t_1} = \frac{V_2}{t_2}$$

With that same three cubic feet of gas at 68 °F, what will be the volume at 212 °F? Because this law is valid only with absolute temperatures (no negative numbers allowed), we must first convert the Fahrenheit numbers to absolute. With Fahrenheit temperatures the easiest conversion is to the Rankine Scale by adding 460. Therefore:

$$P_1 = P_2 \qquad V_1 = 3 \qquad t_1 = 68 + 460 =$$

$$P_2 = P_1 \qquad V_2 = ? \qquad t_2 = 212 + 460 =$$

Applying Charles's law:

$$\frac{V_1}{t_1} = \frac{V_2}{t_2}$$

$$\frac{3}{528} = \frac{V_2}{672}$$

Simplify by cross-multiplying:

$$3 \times 672 = 2016$$

$$V_2 \times 528 = 528\ V_2$$

Therefore: $$528\ V_2 = 2016$$

and: $$V_2 = \frac{2016}{528}$$

or: $$V_2 = 3.8\ \text{cubic feet}$$

Rarely in real life will a firefighter come up against a situation wherein the pressure is held constant while the volume and temperature changes, or the temperature is constant while the pressure and volume changes. More often than not, a firefighter must be concerned with a container of gas (or evaporating liquid) in an enveloping fire—in gas law terminology, a constant volume with increasing temperature. The hazard arises, in this situation, as the pressure increases to the failure point of the tank or container.

This is described by the gas law which states that the pressure of a given volume of gas is directly proportional to the absolute temperature, as long as the volume does not change. This is known as Gay-Lussac's law. It is expressed mathematically as:

$$\frac{P_1}{t_1} = \frac{P_2}{t_2}$$

example: If a container of gas at 900 psi and 80 °F is in enveloping fire and the temperature could rise to 1600 °F, what would the resultant pressure be?

In a container, before it bursts, V_1 is equal to V_2. Therefore:

$$V_1 = V_2 \qquad P_1 = 900 \qquad t_1 = 80 + 460$$

$$V_2 = V_1 \qquad P_2 = ? \qquad t_2 = 1600 + 460$$

Applying Gay-Lussac's law:

$$\frac{P_1}{t_1} = \frac{P_2}{t_2}$$

$$\frac{900}{540} = \frac{P_2}{2060}$$

Simplify by cross-multiplying:

$$900 \times 2060 = 1854000$$

$$540 \times P_2 = 540P_2$$

Therefore: $540P_2 = 1854000$

and: $$P_2 = \frac{1854000}{540}$$

or: $P_2 = 3433$ psi

TRANSITION

At normal temperature and pressure, all substances exist in one of the three states of matter, but many of them can and do exist in one or more of the other two states, depending on whether the temperature and pressure of their surroundings is higher or lower than normal. Normal temperature and pressure is called room temperature and pressure and is set at 68 °F (20 °C) and 1 atmosphere, 14.7 psi (101.35 kilopascals).

Most solids will become a liquid when brought to a high enough temperature. The transition temperature between solid and liquid states is called the **melting point** of the substance. If the heat input continues, most liquids will become a gas at some other definite temperature. This transition temperature between the liquid and gaseous states is called the **boiling point**. Water is normally a liquid, but below 32 °F (0 °C) it becomes a solid (ice). The **freezing point** is, therefore, essentially the same as the melting point. It is not always exactly the same but it is close enough for our purposes. The name given to the particular action indicates the direction of the transition: from liquid to solid is called the freezing point while from solid to liquid is called the melting point.

Because of the frenzy of molecular activity going on within them, liquids tend to become gases even at temperatures well below their boiling points. This transition is slow but continual, and is called **evaporation**. The property of the liquid which is a measure of that tendency is called its **vapor pressure**.

While gases and liquids behave in fairly predictable fashion, not all solids go through the progressive transition points from solid to liquid to gas. Some solids will skip the liquid state and go directly into the gaseous state at the melting point. When a solid does this, it is said to **sublime**. While still the melting point, it is more correctly referred to as the **sublimation point**.

PROPERTIES

While sublimation, melting, and boiling points are very definite and precise properties of materials, they are but a few of the many that chemists use to describe the multitude of substances existing in this world. When taken as a whole, these collective characteristics describe one and only one substance. They are the fingerprints of the chemical world. These properties are divided into two types, physical and chemical.

Physical properties are those which are inherent to the substance, such as color, form, texture, taste, and, in some cases, smell. They are all readily discernable to any observer. More subtle yet still physical properties include boiling point, melting point, viscosity, and density. The determination of these fingerprints requires more training and often more than average everyday instrumentation.

DENSITY AND SPECIFIC GRAVITY

Density is that physical property which is a measure of the mass of a unit volume of the material—solid, liquid, or gas. It is usually expressed as so many grams per cubic meter, or centimeter, for solids and liquids, and in grams per liter for gases.

A related term is **specific gravity**. This term expresses the ratio of the mass of a body to the mass of an equal body of water at 4 °C or some other temperature which must then be specified.

The mass can be expressed in pounds, grams, gallons, tons, or any other unit, as long as the same quantity of water is used in the ratio. Because it is a ratio, specific gravity is always expressed as a number alone without any identifying terminology. Because the mass of one cubic centimeter of water at 4 °C is one gram, specific gravity and density are numerically equal at this reference temperature. They are *not,* however, the same term.

A familiar example of specific gravity is that of the evaluation of the solution in a lead-acid storage battery. The specific gravity of this sulfuric acid and water solution decreases as the battery goes from full charge to discharge, and increases again as the battery is recharged. In a discharged battery, the liquid (acid) contains less hydrogen sulfate and more water molecules than a battery which is charged. With this increase in water and decrease in hydrogen sulfate,

Density of Various Substances

Substance	Density (grams/cubic centimeter)
concrete	2.7–3.0
diamond	3.01–3.52
glass	2.4–2.8
paper	0.7–1.15
ice	0.917
sugar	1.59
ethyl alcohol	0.791
mercury	13.6
gasoline	0.66–0.69

the specific gravity of the solution approaches 1.0, which is the specific gravity of water. This gain and loss of water is illustrated by the chemical reaction which takes place:

$$PbO_2 + Pb + 2H_2SO_4 \xrightleftharpoons[\text{Discharge}]{\text{Charge}} 2PbSO_4 + 2H_2O + \text{Energy}$$

The specific gravity of a fully charged lead-acid battery would be about 1.21, while a completely discharged battery would be about 1.16. The specific gravity is increased by the endothermic reaction brought about by the energy input from the charger as it returns the battery to full charge.

VISCOSITY

Another property specific to liquids and important to fire safety personnel is viscosity. Thin liquids are said to have low viscosity, thick or sirupy liquids to have high viscosity.

Viscosity is a very complex subject but basically concerned with the intermolecular forces of fluids. The study of viscosity is called **rheology**. Simply stated, viscosity is the resistance of a liquid to flow over itself. An event that is similar to viscosity is the beginning of the ice floes in the Niagara or upper Mississippi rivers in the spring. The resistance to the chunks of ice going downstream is proportional to the size of the pieces. The larger the pieces of ice, the more resistance there is to the pieces flowing up and over each other.

So in a liquid, the greater the intermolecular forces, the more the liquid begins to resemble a solid and resists flow, and the higher the viscosity.

There are two basic types of viscosity or flow characteristics, called Newtonian and non-Newtonian. A Newtonian flow is simple flow that changes only with temperature; molasses is a good example of this type. Non-Newtonian flow is quite another story. In this case the viscosity varies indirectly with the amount of work done on the liquid. In other words the viscosity of a non-Newtonian liquid will become less as it is stirred; the liquid becomes thinner. When the work (energy input) is stopped, the liquid will usually return to its original viscosity after some finite time, which varies with different liquids. Mayonnaise is an example of a non-Newtonian liquid.

SOLUTIONS

In our everyday lives, we do not often deal with pure elements. The substances we use daily are usually mixtures of elements or solutions. A solution is defined as two or more substances in a homogeneous mixture. However, a homogeneous mixture is *not* always a solution. A solution is said to vary within limits, while a homogeneous mixture has no limits.

A solution may be made up of gases, liquids, or solids, or any combination of the three states of matter. To the chemist, a solution need not be a liquid, but it may be made up of two solids.

When two or more metallic solids are made into a homogeneous mixture, we commonly call it an **alloy.** However, not all alloys are solutions. In metallic nonsolutions, the components either retain their own physical characteristics, such as crystalline structure, or form new intermetallic material.

In a solution, one of the components will normally be a greater part of the whole than the other component or components. The component that constitutes that greater part is called the **solvent,** and the lesser component (or components) is called the **solute.** A help in remembering which is which: the word solvent is larger, having seven letters, and the solvent is always the larger percentage of the solution.

When two or more gases are mixed together, there is no limit to the proportions. Gases mix in all proportions because of their relatively low densities and the great spaces between their atoms or molecules. Two or more solids can be mixed in *almost* any proportion, but liquids present a much more complex problem in some mixtures. Some liquids will not stay mixed unless special homogenizing components or techniques are used. Even then, the resulting mixture is not always a homogeneous solution in the strict sense.

Gases will dissolve in liquids, but only in very definite and characteristic amounts.

When solids are put into liquids, once again, there is a very definite and characteristic amount that can go into solution at any one temperature. Temperature is very important; the higher the temperature, the more solid will go into solution.

In these solid-liquid solutions, two mechanisms operate causing the solid to disappear and thus form the homogeneous mixture. When mixed with water, some solids will conduct an electric current, while others will not. Those that conduct electricity are called **electrolytes;** those that do not are called **nonelectrolytes.**

Sugar is a nonelectrolyte and, as such, will not conduct an electric current when the sugar is dissolved in water. The sugar molecule retains its identity surrounded by water molecules.

When table salt is added to water, the solution mechanism is complex, but it can be simplified and visualized thus: the salt is the element sodium and the element chlorine tied together by sharing an electron for every molecule formed. When the sodium chloride molecule is placed into the water, the water acts like a knife that causes the separation of the sodium chloride molecule into a sodium ion, a mass with a net positive charge, and a chlorine ion, a mass with a net negative charge. This is *not* to say the sodium is present in the water in its elemental form with all of its violent antiwater characteristics, or that chlorine gas will rise from the water. The sodium chloride is still present in the water and retains all of its own unique properties. If, however, we were to supply energy in the form of free electrons (such as an electric current) to the solution by connecting metal plates to wires and the wires to a battery, after immersing the plates in the solution we would see chlorine gas bubbles rise from the plate connected to the positive terminal of the battery and metallic sodium beginning to collect at the negative plate. This reaction occurs because of the unbalance or net charge property of the ions and their desire to follow that law of nature to be balanced. Sodium atoms collect at the negative plate because the plate possesses an excess of electrons. Each sodium atom picks up an electron to become neutral and elemental. The chlorine ions, with the excess electrons, discharge them at the positive plate where there is a lack of electrons. The chlorine ion therefore becomes a chlorine atom with all the properties of elemental chlorine. This transformation of sodium chloride into the elements sodium and chloride is an endothermic reaction. It will stop as soon as the input energy from the battery is removed.

These electrolytes and nonelectrolytes have a very definite influence on the freezing and boiling points of liquids. The boiling and freezing points are changed according to the amount of solute in the solvent. The freezing point is lowered and the boiling point is raised as the solute is added to the solution. Both of these changes depend not only on the concentration and nature of the solute, but the concentration and nature of the solvent as well. These changes are exactly the characteristics that make antifreezes or summer coolants do their job. It is also the reason why one compound can raise the boiling point, as well as lower the freezing point.

REVIEW QUESTIONS

1. Write a paragraph explaining the basic concept of the construction of the atom.
2. Why do we postulate the mass called a neutron?
3. Explain the periodic law.

4. Using the symbols A and B write:
   a. a composition reaction
   b. a decomposition reaction
   c. a simple replacement reaction
   d. a double displacement reaction
5. What are valence electrons? Explain their role in chemical reactions.
6. Explain polyvalency. Give an example.
7. How do compounds differ from atoms and molecules?
8. What is oxidation? What is reduction? Explain both in detail.
9. What is the difference between physical and chemical changes? Give examples of both and label each.
10. Distinguish between potential and kinetic energy.
11. Explain the difference between mass and weight.
12. What is an endothermic reaction? What is an exothermic reaction? Give several examples of each.
13. Temperature and heat are related. What is the exact difference between them?
14. What is a degree? What is a calorie? How many calories in a Kcal?
15. List the states of matter and explain the difference between them.
16. Explain melting point, boiling point, freezing point, and sublimation.
17. If the original pressure on a cylinder is 3 atm and the volume of that cylinder is 6 cubic feet, what would the resulting pressure be if the volume is compressed very slowly to 1 cubic foot?
18. Using an accurate pressure relief valve to keep the pressure constant, what volume change will occur in a 10 cubic foot cylinder when the temperature is increased from 70 °F to 200 °F?
19. What is the distinction between density and specific gravity?
20. In a solution, which is the solvent and which is the solute?
21. Give the names of the elements related to the following chemical symbols: Na, Pb, Hg, Ag, S, Au, N, Al, C, Bk, Cf, Cu, H, and K.

Basic Organic Chemistry/ 3

SIMILARITIES AND DIFFERENCES

Chemistry is usually divided into two basic areas, organic and inorganic chemistry. There are also many subdivisions in each area.

Inorganic chemistry concerns itself with the properties and reactions of all elements and their compounds with the exception of most of the compounds of carbon. Organic chemistry concerns itself with the properties and reactions of the compounds of carbon. The number of those compounds covered by organic chemistry is tremendous and is growing almost daily.

All of the laws, theories, and definitions which we have discussed in the preceding chapters are valid for both inorganic and organic chemistry. Inorganic chemistry is usually considered to be the less complex of the two fields because: inorganic reactions usually take place between two different elements; the molecules are usually smaller in size; and, in general, the reactions are considerably more limited than in organic chemistry. The basic reason for the study of the single element carbon being segregated into a unique branch, organic chemistry, is the carbon's ability to unite with other carbon atoms in chemical reactions. This leads to the formation of long chains, long chains with branches, or even closed loops made up of atoms, formed by carbon atom linking to carbon atom. The result is an almost infinite number of possible combinations.

These compounds, long suspected of being different, were originally associated only with life processes. The study of these life processes and the substances connected with them was then called life chemistry. In the late 1820s, a German chemist named Whöler produced a chain of carbon atoms with hydrogen, nitrogen, and oxygen atoms attached to the carbon. It was only a two carbon chain but nevertheless it was the first synthesis of a life chemistry compound. It is known today as urea. There are now over 500,000 compounds of carbon known, more than ten times that known for all other elements combined.

Because the molecules formed by carbon to carbon linking can attain immense size and complexity, the formulas for organic chemistry require additional shorthand techniques. Nicotine, that infamous compound in cigarette smoke, would be written $C_{10}H_{14}N_2$ and read the same way inorganic formulas are read. However, the formula is really an oversimplification. It does not tell us the true relationships or relative positions of the individual atoms. For this reason we must add a diagraming technique. The diagraming technique most often used is named after a man named Kekulé (Keck-u-lay) and was developed in about 1880.

The Kekulé diagram for nicotine is shown in Figure 3.1. Counting the carbons (C), the nitrogens (N), and the hydrogens (H) it will be found that there are ten carbons, fourteen hydrogens, and two nitrogens in the molecule ($C_{10}H_{14}N_2$).

Another material familiar to every firefighter is nylon. While there are a whole family of nylon molecules, the basic formula for nylon is $[NH(CH_2)_4NHCO(CH_2)6CO]_x$. This means that nylon is a chain of some number (x) of molecules made up of repeating atoms of nitrogen and carbon with either hydrogen atoms or oxygen atoms together with groups of carbon and hydrogen atoms arrayed off the main chain. These groups arrayed off the main chain are called **branches.**

Nylon would have the Kekulé diagram seen in Figure 3.2.

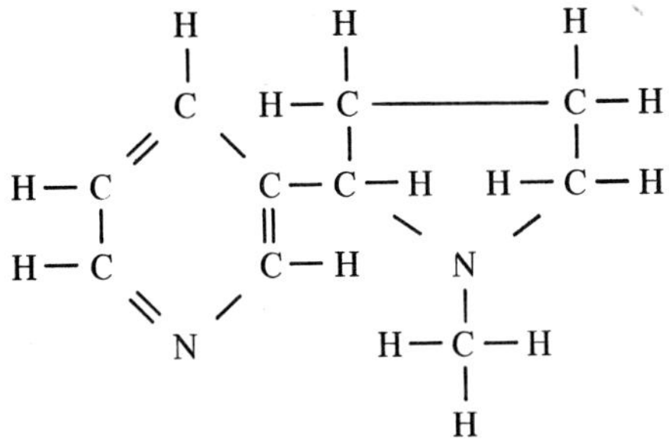

Figure 3.1. Nicotine.

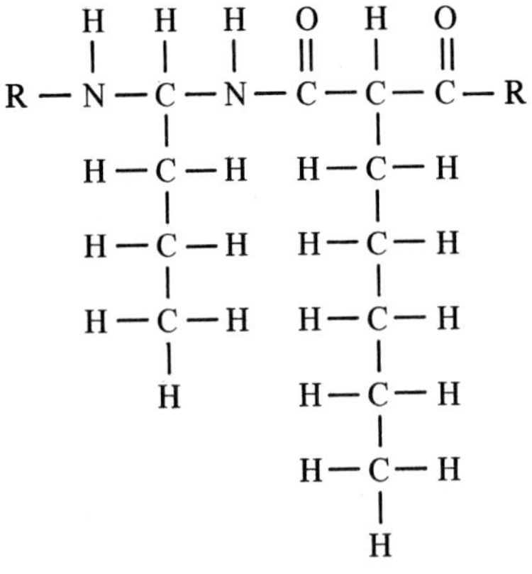

Figure 3.2. Nylon. R Indicates a Repeat of the Same Molecule.

ALIPHATIC COMPOUNDS

Nylon and nicotine belong to two different families of organic compounds. Nylon is classed as an **aliphatic** while nicotine belongs to a group called **aromatics**.

Aliphatic compounds are characteristically compounds with straight carbon chains, branches, and nonresonant ring structures. Aromatic compounds will be discussed later.

The aliphatic compounds are further subdivided into **saturated** and **unsaturated** series. An aliphatic molecule is considered saturated if the links between the carbon atoms are all single valence bonds. For example, ethane looks like this:

$$\begin{array}{ccc} & H & H \\ & | & | \\ H - & C - C & - H \\ & | & | \\ & H & H \end{array}$$

Because all valence bonds are filled the chemical reactivity of this series is relatively low, and the series is also called the **paraffin series** from the Latin "parum" (little) and "affinis" (affinity). They are also known as **alkanes**.

If the linkages between any two carbon atoms in a molecule are double or triple, the compound is considered unsaturated:

$$\begin{array}{ccc} & H & H \\ & | & | \\ H - & C = C & - H \\ & | & | \\ & H & H \end{array}$$ ethylene (alkene series)

$$\begin{array}{ccc} & H & H \\ & | & | \\ H - & C \equiv C & - H \\ & | & | \\ & H & H \end{array}$$ acetylene (alkyne series)

With carbon chains longer than two carbon atoms, combinations of saturated and unsaturated bonds may exist. The total compound, however, is considered unsaturated as long as at least one of the carbon-to-carbon linkages is unsaturated.

Figure 3.3 shows the relationships of a few of the saturated and unsaturated aliphatic hydrocarbons. The complete series cannot be shown because the total is practically unlimited.

Number of Carbons	Saturated			Unsaturated (double bond)			Unsaturated (triple bond)		
	Name	Formula	Carbon Linkage	Name	Formula	Carbon Linkage	Name	Formula	Carbon Linkage
1	Methane	CH_4	—C—						
2	Ethane	C_2H_6	C—C	Ethylene	C_2H_4	C=C	Acetylene	C_2H_2	C≡C
3	Propane	C_3H_8	C—C—C	Propylene	C_3H_6	C=C—C	Methylacetylene	C_3H_4	C—C≡C
4	Butane	C_4H_{10}	C—C—C—C	Butene-1	C_4H_8	C=C—C—C	Ethylacetylene	C_4H_6	C—C—C≡C
				Butene-2	C_4H_8	C—C=C—C	Dimethylacetylene	C_4H_6	C—C≡C—C
5	Pentane	C_5H_{12}	C—C—C—C—C	Pentene-1	C_5H_{10}	C=C—C—C—C	Pentine-1	C_5H_8	C≡C—C—C—C
				Pentene-2	C_5H_{10}	C—C=C—C—C	Pentine-2	C_5H_8	C—C≡C—C—C
6	Hexane	C_6H_{14}	C—C—C—C—C—C	Hexene-1	C_6H_{12}	C=C—C—C—C—C	Hexine-1	C_6H_{10}	C≡C—C—C—C—C
				Hexene-2	C_6H_{12}	C—C=C—C—C—C	Hexine-2	C_6H_{10}	C—C≡C—C—C—C
				Hexene-3	C_6H_{12}	C—C—C=C—C—C	Hexine-3	C_6H_{10}	C—C—C≡C—C—C
7	Heptane	C_7H_{16}	C—C—C—C—C—C—C	Heptene-1	C_7H_{14}	C=C—C—C—C—C—C	Heptine-1	C_7H_{10}	C≡C—C—C—C—C—C
8	Octane	C_8H_{18}	C—C—C—C—C—C—C—C	Octene-1	C_8H_{16}	C=C—C—C—C—C—C—C	Octine-1	C_8H_{12}	C≡C—C—C—C—C—C—C

Figure 3.3. Carbon-to-Carbon Relationships of a Few of the Basic Saturated and Unsaturated Aliphatic Hydrocarbons.

ISOMERS

With one, two, or three carbon chains, there is only one logical Kekulé structure: a straight chain or a highly stressed ring. However, with four carbons, alternates begin to appear.

Butane normally has the structure:

$$
\begin{array}{c}
\quad\text{H}\quad\text{H}\quad\text{H}\quad\text{H} \\
\quad|\quad\ |\quad\ |\quad\ | \\
\text{H}-\text{C}-\text{C}-\text{C}-\text{C}-\text{H} \qquad C_4H_{10} \\
\quad|\quad\ |\quad\ |\quad\ | \\
\quad\text{H}\quad\text{H}\quad\text{H}\quad\text{H}
\end{array}
$$

The following structure also exists:

$$
\begin{array}{c}
\text{H} \\
| \\
\text{H}-\text{C}-\text{H} \\
| \\
\text{H}\quad\ |\quad\ \text{H} \\
|\quad\ |\quad\ | \\
\text{H}-\text{C}-\text{C}-\text{C}-\text{H} \qquad C_4H_{10} \\
|\quad\ |\quad\ | \\
\text{H}\quad\text{H}\quad\text{H}
\end{array}
$$

When both molecules have the same molecular formula (such as C_4H_{10}) but different physical structures, they are called isomers. The name is derived from "iso-" meaning "the same" and "-mer" meaning "smallest complete part." The first molecule of the above example is called normal butane or n-butane, and the second, iso-butane. The word normal indicates the basic or unbranched chain isomer.

With pentane, 3 isomers are possible:

$$
\begin{array}{c}
\text{H}\quad\text{H}\quad\text{H}\quad\text{H}\quad\text{H} \\
|\quad\ |\quad\ |\quad\ |\quad\ | \\
\text{H}-\text{C}-\text{C}-\text{C}-\text{C}-\text{C}-\text{H} \qquad C_5H_{12}\quad \text{n-pentane} \\
|\quad\ |\quad\ |\quad\ |\quad\ | \\
\text{H}\quad\text{H}\quad\text{H}\quad\text{H}\quad\text{H}
\end{array}
$$

$$
\begin{array}{c}
\text{H} \\
| \\
\text{H}-\text{C}-\text{H} \\
| \\
\text{H}\quad\text{H}\quad\ |\quad\ \text{H} \\
|\quad\ |\quad\ |\quad\ | \\
\text{H}-\text{C}-\text{C}-\text{C}-\text{C}-\text{H} \qquad C_5H_{12}\quad \text{iso-pentane} \\
|\quad\ |\quad\ |\quad\ | \\
\text{H}\quad\text{H}\quad\text{H}\quad\text{H}
\end{array}
$$

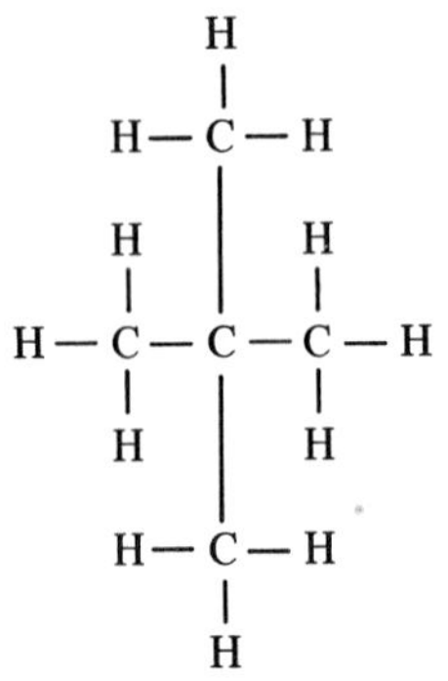

```
          H
          |
      H — C — H
          |
      H   |   H
      |   |   |
  H — C — C — C — H      C₅H₁₂   neo-pentane
      |   |   |
      H   |   H
          |
      H — C — H
          |
          H
```

Realize that the structure

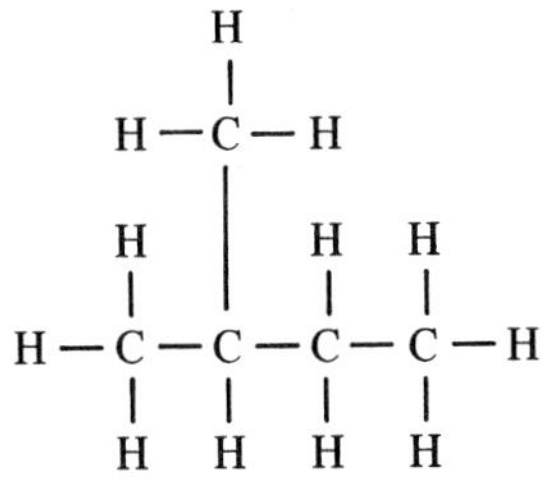

```
              H
              |
          H — C — H
              |
      H   |   H   H
      |   |   |   |
  H — C — C — C — C — H
      |   |   |   |
      H   H   H   H
```

is no different than the second structure shown above. It is merely a mirror image of it. It would still be the same molecule if the branch projected downward instead of upward.

As the number of carbons in the chain increases, the number of possible isomers increases. We have seen that with a four carbon chain two isomers are possible; with five in the chain, three isomers may exist. With six carbons in the chain five isomers are possible, and so on. Someone has calculated that $C_{14}H_{30}$ may have 1858 isomers and $C_{40}H_{82}$ over 62 trillion possible isomers.

Because of the existence of branches, additional rules are used in naming the isomers:

- The longest chain of carbon atoms is used to give the molecule its name.
- The carbons in this principal chain are numbered from left to right so that the carbon atom where the branch exists can be noted.
- The branches are named according to their length just as are the principal chains, with the lowest numbered branch first, where multiple branching exists. Branches are said to be made up of radicals using a -yl suffix. If the branches are not chains but simple elements, they too are named by noting the atoms to which they are attached:

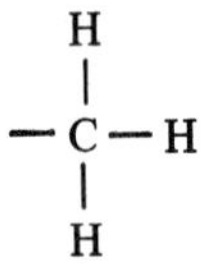

```
          H
          |
      — C — H        methyl radical
          |
          H
```

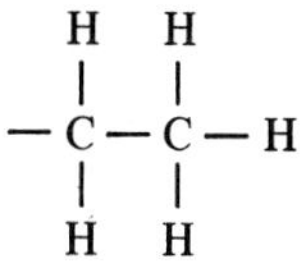

H H
| |
— C — C — H   ethyl radical
| |
H H

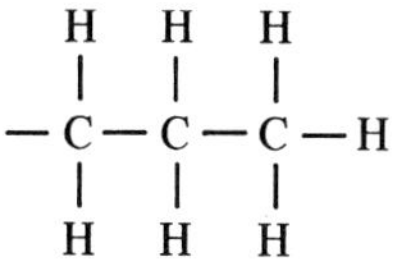

H H H
| | |
— C — C — C — H   propyl radical
| | |
H H H

The name of the compound is then a combination of the branch name or names followed by the principal chain designation.

example:

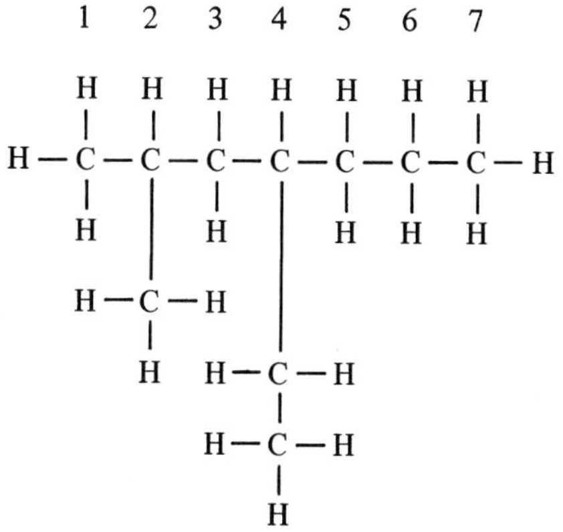

1 2 3 4 5 6 7

H H H H H H H
| | | | | | |
H — C — C — C — C — C — C — C — H   (The length of the linkage
| | | | | | has no significance.)
H | H | H H H

H — C — H
|
H   H — C — H
|
H — C — H
|
H

1. Branches are named first according to position and length: 2-methyl, 4-ethyl.
2. Then the principal chain is named: heptane
3. Adding it all together, the name of the compound is: 2-methyl,4-ethylheptane

example:

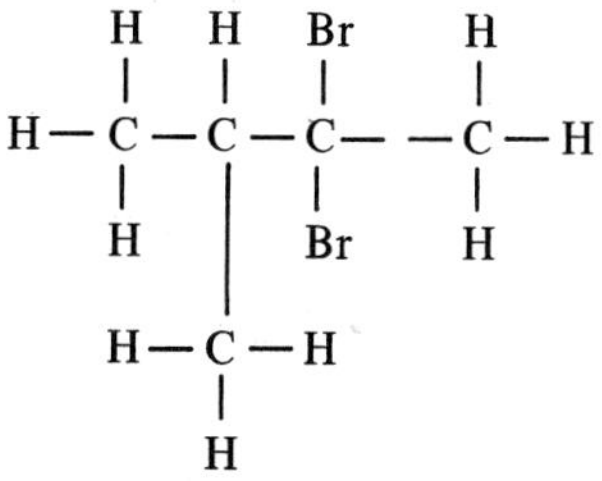

H H Br H
| | | |
H — C — C — C— —C — H
| | | |
H | Br H

H — C — H
|
H

1. Again, branches are named first: 2 methyl, 3,3 dibromo (di=two)
2. The principal chain is named: butane
3. The proper name of the compound is: 2 methyl 3,3 dibromobutane

- When the principal carbon chain or one of the branches is unsaturated, an additional number is added after the unsaturated designation to indicate where the unsaturation is located.

example:

$$
\begin{array}{cccc}
\text{H} & \text{H} & \text{H} & \text{H} \\
| & | & | & | \\
\text{H}-\text{C}=\text{C}-\text{C}-\text{C}-\text{H} \\
& & | \\
& & \text{H} \\
\\
& \text{H}-\text{C}-\text{H} \\
& & | \\
& & \text{H}
\end{array}
$$

1. Branches are named first: 2 methyl
2. Then the principal chain is named: butene (-ene=unsaturated)
3. The position of double bonds is designated: carbon 1
4. So the proper name is: 2-methylbutene-1

Aliphatic ring structures can also be saturated or unsaturated. When, for example, pentane forms a ring rather than a straight chain, it is called cyclopentane and has the structure

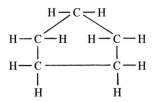

Here also groups can be attached to the ring on one or more of the carbons in the ring.

AROMATIC SERIES

The parent hydrocarbon ring of the greatest number of organic cyclic compounds is a structure with six carbons and six hydrogens, indicating that unsaturation or double bonds must exist in those molecules. It has the Kekulé structure

$$
\begin{array}{c}
\text{H} \\
| \\
\text{C} \\
\text{H}-\text{C} \quad \text{C}-\text{H} \\
\text{H}-\text{C} \quad \text{C}-\text{H} \\
\text{C} \\
| \\
\text{H}
\end{array}
$$

After studying the naming of compounds, and realizing that ring structures are numbered starting at 12 o'clock, the temptation is to give the above structure the name hexene-1,3,5, indicating double bonds at the first, third, and fifth bond positions. However, studies have shown that the bonds in this molecule are so highly dynamic that the structure is called a **resonant hybrid**. This means that it acts as though both forms exist simultaneously, thus:

$$
\begin{array}{c}
\text{H} \\
| \\
\text{C} \\
\text{H}-\text{C} \quad \text{C}-\text{H} \\
\text{H}-\text{C} \quad \text{C}-\text{H} \\
\text{C} \\
| \\
\text{H}
\end{array}
\rightleftharpoons
\begin{array}{c}
\text{H} \\
| \\
\text{C} \\
\text{H}-\text{C} \quad \text{C}-\text{H} \\
\text{H}-\text{C} \quad \text{C}-\text{H} \\
\text{C} \\
| \\
\text{H}
\end{array}
$$

This molecule is commonly called **benzene** or the **benzene ring** and is the base molecule for the aromatic series.

It was because the first identification of the benzene ring-based compounds was with substances which produced aroma that the benzene family or series has become known as the aromatic series. Although we now know that not all benzene ring-based compounds have an aroma, we still classify all benzene ring derivatives as aromatics.

Many chemists, when writing the benzene ring, in order to save time, will use the symbol

understanding that all six carbons are at the corners, the six hydrogens are attached to the carbons, and the double bonds resonate. We shall use this technique in this book.

While benzene is the base for all the aromatics, there are many subgroups with the benzene ring structure. These groups have a number of benzene rings linked together, and can form very large and complicated molecules.

A popular marine cargo is naphthalene, which is shipped as both a solid and a liquid. Naphthalene has the double benzene ring structure

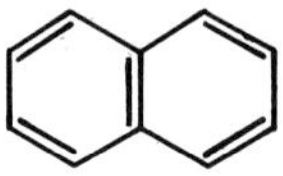

naphthalene

Two of the other subseries are:

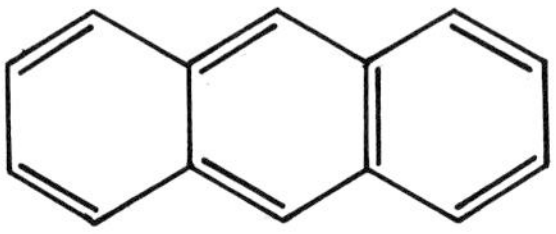

anthracene

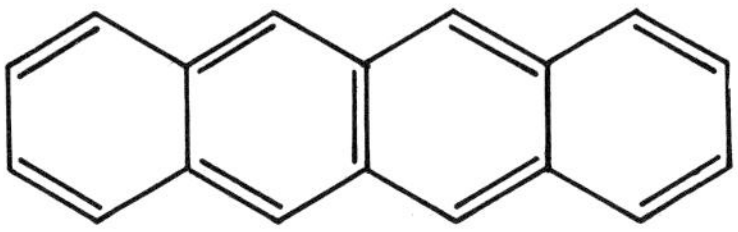

naphthacene

The possible combinations of benzene rings with each other or with other elements are almost infinite. Just about any element that can be made to react with carbon can be attached to the ring or rings to tailor-make molecules with almost any desired properties. For example, adding a methyl ($-CH_3$) group to a benzene ring by replacing one hydrogen forms toluene:

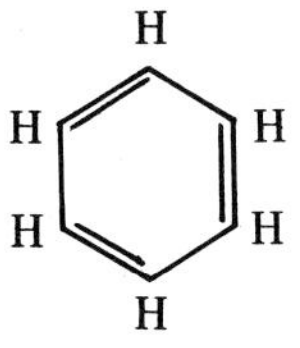

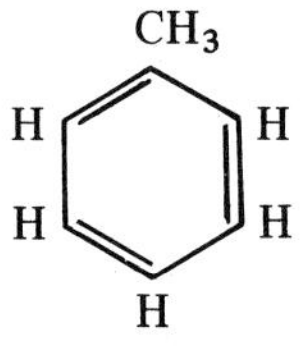

Adding a second methyl group by replacing a second hydrogen leads to three different possibilities depending upon which hydrogen is replaced.

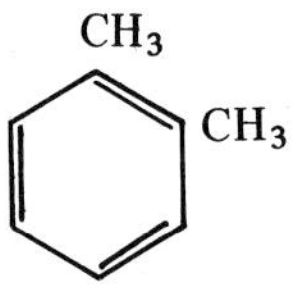

1. 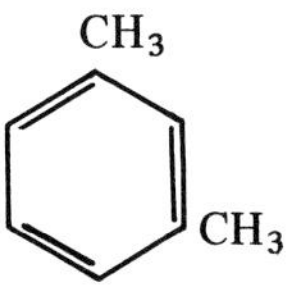   2.    3.

In benzene ring nomenclature, the carbons are numbered clockwise with number one at the 12 o'clock position. Molecule one, above, would therefore be known as 1,2 dimethylbenzene. Number two would be 1,3, dimethylbenzene, and number three would be 1,4 dimethylbenzene. We know a mixture of all three compounds by the common name **xylene.**

Obviously xylene is made up of three isomers, commonly called *ortho*-xylene (groups on adjacent carbon atoms), *meta*-xylene (groups separated by one carbon atom), and *para*-xylene (groups separated by two carbon atoms). Ortho is usually abbreviated *o*-, meta *m*-, and para *p*-.

The methyl group of the above xylene example might be any other group, element, or ring compound. Cresol isomers, for example, have the Kekule structures:

| ortho-cresol | meta-cresol | para-cresol |
| o-cresol | m-cresol | p-cresol |

Creosote oil contains a large percentage of these three isomers in a mixture.

As with the aliphatics, while certain aromatic structures may look different at first glance, they may be the same. An ortho compound would be the same molecule whether the second radical or group were connected to the number two or to the number six carbon. These molecules would be mirror images. So also would those molecules with a second radical or group on the three and five carbon atom.

ORGANIC CHEMICAL FAMILIES

As human families have family characteristics so, too, do chemical families. Each compound of the family has a unique group or radical attached to the molecule, or there are two or more groups united by a unique linkage. A few of the more important families for fire safety personnel are:

Family Name	Typical Suffix	Characteristic	Example	Specific Name
alcohol	-anol	$-C-O-H$	$H-C-OH$	methanol
aldehyde	-al	$-C=O$	$Cl-C-C=O$	chloral
amine	-amine	$-C-N-H$	$H-C-N-H$	ethylamine

Family Name	Typical Suffix	Characteristic	Example	Specific Name
ester	(variable)	$-\overset{\overset{O}{\|\|}}{C}-O-$	$H-\overset{\overset{H}{\|}}{\underset{\underset{H}{\|}}{C}}-\overset{\overset{O}{\|\|}}{C}-O-\overset{\overset{H}{\|}}{\underset{\underset{H}{\|}}{C}}-\overset{\overset{H}{\|}}{\underset{\underset{H}{\|}}{C}}-H$	ethyl acetate
ether	ether	$-C-O-C-$	$H-\overset{\overset{H}{\|}}{\underset{\underset{H}{\|}}{C}}-\overset{\overset{H}{\|}}{\underset{\underset{H}{\|}}{C}}-O-\overset{\overset{H}{\|}}{\underset{\underset{H}{\|}}{C}}-\overset{\overset{H}{\|}}{\underset{\underset{H}{\|}}{C}}-H$	diethyl ether
ketone	-one	$-\overset{\overset{O}{\|\|}}{C}-\overset{O}{C}-C-$	$H-\overset{\overset{H}{\|}}{\underset{\underset{H}{\|}}{C}}-\overset{\overset{O}{\|\|}}{C}-\overset{\overset{H}{\|}}{\underset{\underset{H}{\|}}{C}}-H$	acetone
organic acid	acid	$-\overset{\overset{O}{\|\|}}{C}-O-H$	$H-\overset{\overset{H}{\|}}{\underset{\underset{H}{\|}}{C}}-\overset{\overset{O}{\|\|}}{C}-O-H$	acetic acid
peroxide	peroxide	$-C-O-O-C-$	$H-\overset{\overset{H}{\|}}{\underset{\underset{H}{\|}}{C}}-\overset{\overset{H}{\|}}{\underset{\underset{H}{\|}}{C}}-O-O-\overset{\overset{H}{\|}}{\underset{\underset{H}{\|}}{C}}-\overset{\overset{H}{\|}}{\underset{\underset{H}{\|}}{C}}-H$	ethyl peroxide

ALCOHOL

Alcohols are characterized by the hydroxyl group (— OH). Many organic chemists consider straight chain alcohols to be an organic relation or derivative of water. Methyl alcohol has the structure

$$H-\overset{\overset{H}{\|}}{\underset{\underset{H}{\|}}{C}}-OH$$

while water, of course, has the structure $H - OH$.

Early in their history, the then-known alcohols, methyl and ethyl, were derived from the distillation of wood and grains. They therefore received the common names of wood alcohol (methyl) and grain alcohol (ethyl). They are, even today, often called wood or grain spirits.

The straight chain alcohols may be either liquid or solid depending on their molecular weight. With only one exception, from methyl alcohol (CH_3OH) up to lauryl alcohol ($C_{12}H_{25}OH$) they are liquids. (Lauryl and decyl [$C_{10}H_{21}OH$] are thick, oily liquids.) The one exception is tertiary butyl alcohol which has the structure:

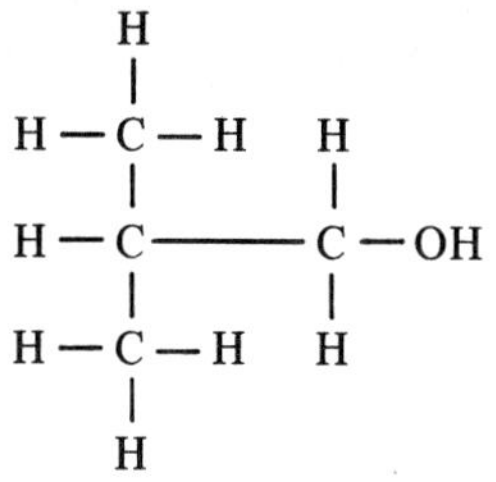

$$
\begin{array}{c}
\text{H} \\
| \\
\text{H}-\text{C}-\text{H} \quad \text{H} \\
| \qquad\qquad | \\
\text{H}-\text{C}\text{——}\text{C}-\text{OH} \\
| \qquad\qquad | \\
\text{H}-\text{C}-\text{H} \quad \text{H} \\
| \\
\text{H}
\end{array}
$$

This solid alcohol has a melting point of 78 °F (26 °C). Above lauryl the straight chain alcohols are all solids.

Of the many aromatic alcohols, phenol, also known as carbolic acid, is perhaps the most common. It has the family trait (—OH) in its structure, thus:

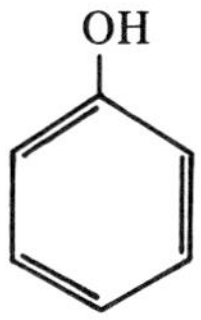

OH

Phenol is a solid with a melting point of 109 °F (43 °C) and a boiling point of 358 °F (181 °C).

Cresol is another common aromatic alcohol with ortho, meta, and para forms:

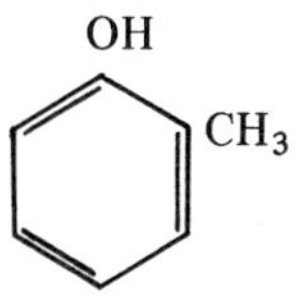

OH	OH	OH
CH$_3$	CH$_3$	CH$_3$
ortho-	meta-	para-

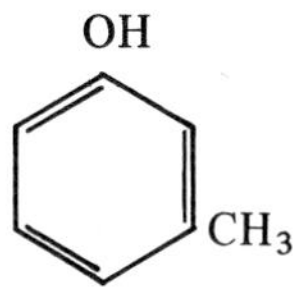

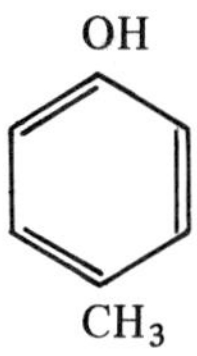

There is also an alpha (*a*) and beta (*β*) form of the aromatic alcohol, naphthol:

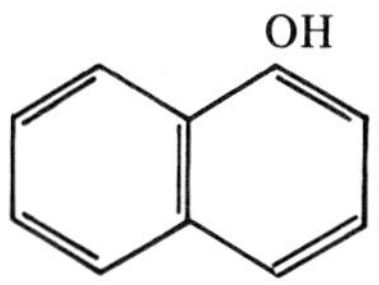

OH

a-naphthol   *β*-naphthol

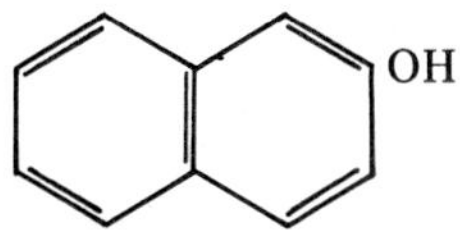

The *a* and *β* detail the relationship of the hydroxyl group to the junction of the cyclic rings.

While the aromatic alcohols have relatively high melting and boiling points, the straight chain alcohols are quite flammable with flash points in the 55 to 60 °F range (13 to 16 °C). They burn with a blue flame.

The straight chain alcohols have slightly higher densities than the corresponding nonhydroxyl organic compounds but nevertheless are usually lighter than water. Because the alcohol is partially oxidized by the addition of that — OH group, the exothermic energy produced by burning alcohols is less than that of the corresponding normal organic compound.

GLYCOLS

If, in a straight chain molecule, not one OH group but *two* are added the result is called a glycol. Thus ethylene glycol would be

$$HO-\underset{\underset{H}{|}}{\overset{\overset{H}{|}}{C}}-\underset{\underset{H}{|}}{\overset{\overset{H}{|}}{C}}-OH$$

and propylene glycol would be

$$HO-\underset{\underset{H}{|}}{\overset{\overset{H}{|}}{C}}-\underset{\underset{H}{|}}{\overset{\overset{H}{|}}{C}}-\underset{\underset{H}{|}}{\overset{\overset{H}{|}}{C}}-OH$$

The addition of the second hydroxyl group is further oxidation of the molecule and increases the boiling point considerably:

no hydroxyl group	ethane $H-\underset{\underset{H}{|}}{\overset{\overset{H}{|}}{C}}-\underset{\underset{H}{|}}{\overset{\overset{H}{|}}{C}}-H$	boiling point −126 °F (−88 °C)
one hydroxyl group	ethanol $H-\underset{\underset{H}{|}}{\overset{\overset{H}{|}}{C}}-\underset{\underset{H}{|}}{\overset{\overset{H}{|}}{C}}-OH$	boiling point 173 °F (78 °C)
two hydroxyl groups	ethylene glycol $HO-\underset{\underset{H}{|}}{\overset{\overset{H}{|}}{C}}-\underset{\underset{H}{|}}{\overset{\overset{H}{|}}{C}}-OH$	boiling point 389 °F (198 °C)

GLYCEROL

When three hydroxyl groups are added to the paraffinic molecule propane, the result is glycerol:

```
      H   H   H
      |   |   |
 H — C — C — C — H
      |   |   |
      O   O   O
      |   |   |
      H   H   H
```

This is a sweet, syrupy liquid whose boiling point is 554 °F (290 °C). Although it is very stable and quite harmless, it is the substance from which nitroglycerine is formed.

When the hydrogens of the hydroxyl groups of glycerol are replaced by nitrate groups the characteristics of the molecule change quite radically. It becomes nitroglycerine:

```
      H   H   H
      |   |   |
 H — C — C — C — H
      |   |   |
      O   O   O
      |   |   |
 O = N   N   N = O
      |   ‖   |
      O   O   O
```

Nitroglycerine actually belongs to another family of organic compounds called **esters.**

ESTERS

Inorganic acids are marked by their hydrogen content while organic acids contain the carboxyl group:

```
 — C = O
    |
    O
    |
    H
```

It is this group that is responsible for the acidity. The carboxyl group is usually written COOH in order to save space. When the hydrogen in the COOH group is replaced by a hydrocarbon group or groups, the compound then is known as an ester.

Acetic acid can be reacted with ethyl alcohol to form an ester called ethyl acetate:

$$
\underset{\substack{| \\ H}}{\overset{\substack{H \\ |}}{H-C}}-C{=}O \;+\; HO-\underset{\substack{| \\ H}}{\overset{\substack{H \\ |}}{C}}-\underset{\substack{| \\ H}}{\overset{\substack{H \\ |}}{C}}-H \;\longrightarrow\; \underset{\substack{| \\ H}}{\overset{\substack{H \\ |}}{H-C}}-\overset{\substack{O \\ ||}}{C}-O-\underset{\substack{| \\ H}}{\overset{\substack{H \\ |}}{C}}-\underset{\substack{| \\ H}}{\overset{\substack{H \\ |}}{C}}-H \;+\; H-O-H
$$

The ester molecule is characterized by the linkage:

$$
\overset{\overset{\textstyle O}{||}}{-C}-O-
$$

The shorter chain, lower molecular weight, esters are very volatile liquids with fruit-like odors. For example, the aroma of bananas comes from isoamylacetate commonly called banana oil:

$$
\underset{\substack{| \\ H}}{\overset{\substack{H \\ |}}{H-C}}-\overset{\substack{O \\ ||}}{C}-O-\underset{\substack{| \\ H}}{\overset{\substack{H \\ |}}{C}}-\underset{\substack{| \\ H}}{\overset{\substack{H \\ |}}{C}}-\underset{\substack{| \\ H}}{\overset{\substack{H \\ |}}{C}}-\underset{\substack{| \\ H}}{\overset{\substack{H \\ |}}{C}}-H
$$

ETHERS

If the nonlinking oxygen of the ester $\left(-\overset{\overset{\textstyle O}{||}}{C}-\right)$ is not present the compound then becomes two organic molecules linked together by a simple oxygen linkage:

$$
\underset{\substack{| \\ H}}{\overset{\substack{H \\ |}}{H-C}}-O-\underset{\substack{| \\ H}}{\overset{\substack{H \\ |}}{C}}-H \qquad \text{dimethyl ether}
$$

$$
\underset{\substack{| \\ H}}{\overset{\substack{H \\ |}}{H-C}}-\underset{\substack{| \\ H}}{\overset{\substack{H \\ |}}{C}}-O-\underset{\substack{| \\ H}}{\overset{\substack{H \\ |}}{C}}-\underset{\substack{| \\ H}}{\overset{\substack{H \\ |}}{C}}-H \qquad \text{diethyl ether}
$$

$$
\underset{\substack{| \\ H}}{\overset{\substack{H \\ |}}{H-C}}-O-\underset{\substack{| \\ H}}{\overset{\substack{H \\ |}}{C}}-\underset{\substack{| \\ H}}{\overset{\substack{H \\ |}}{C}}-H \qquad \text{methyl ethyl ether}
$$

Also, if the hydrogen of the hydroxyl group on an alcohol is replaced by an organic chain, the parent alcohol becomes an ether.

Ethers boil at very low temperatures compared to alcohols and are very volatile, and therefore highly flammable. Their vapors are generally very heavy compared to air, so that they tend to settle. Because of this and their low flash points ethers catch fire quite easily.

Ethers also will oxidize slowly in air, even at room temperatures, to a white crystalline peroxide. These peroxides are very sensitive to impact and shock once they dry. They will explode with little input energy and great violence.

In their vapor state ethers seem to have an affinity for human body cell surfaces, with the result that they produce artificial sleep.

Compared to alcohols, ethers are chemically inert substances. In fact, they seem to act more like the parent substances (paraffins) than any of the ketones, amines, or other derivatives. These paraffinic characteristics include relatively low boiling points and therefore ethers are important substances for consideration by fire safety personnel.

Twin ether-type compounds exist which are known as *gem*-ethers or gem-diethers. The gem- is derived from the Latin word Gemini meaning "twin." Examination of the structure shows that there are really three chains involved with two of them twins and the third, another straight chain. Acetal is a typical *gem*-diether:

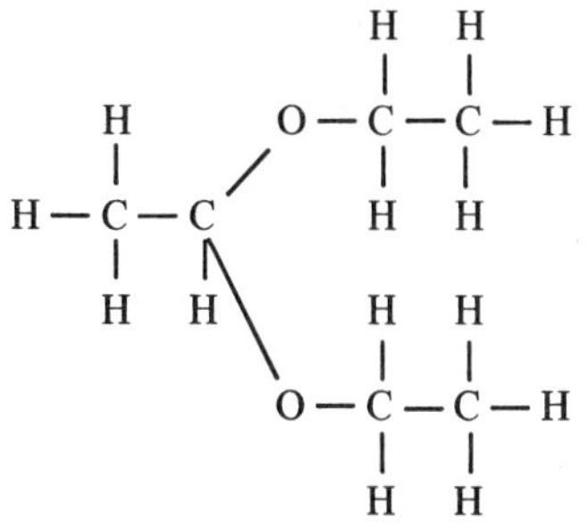

acetal

The boiling point of acetal is 219 °F (104 °C) indicating that the linkage does not radically change that characteristic. The boiling point of a similar ether, dipropyl ether, is 195 °F (91 °C).

FATS, OILS, AND WAXES

A fat may be defined as the ester combination of a special long chain organic acid and glycerol. An oil is merely a fat that is liquid at normal room temperature. Please note that we are not discussing petroleum products; they are not oils in this chemical sense.

The acid parts of fats are often called fatty acids. In natural occurrence, with only one known exception, they always contain an even number of carbon atoms in the molecule:

$$
\begin{array}{c}
\ \ \ \ \ \ H\ \ \ H\ \ \ H\ \ \ O \\
\ \ \ \ \ \ |\ \ \ \ |\ \ \ \ |\ \ \ \ || \\
H - C - C - C - C - OH \\
\ \ \ \ \ \ |\ \ \ \ |\ \ \ \ | \\
\ \ \ \ \ \ H\ \ \ H\ \ \ H
\end{array}
\qquad \text{butyric acid}
$$

$$
\begin{array}{c}
\ \ \ \ \ \ H\ \ \ H\ \ \ H\ \ \ H\ \ \ H\ \ \ O \\
\ \ \ \ \ \ |\ \ \ \ |\ \ \ \ |\ \ \ \ |\ \ \ \ |\ \ \ \ || \\
H - C - C - C - C - C - C - OH \\
\ \ \ \ \ \ |\ \ \ \ |\ \ \ \ |\ \ \ \ |\ \ \ \ | \\
\ \ \ \ \ \ H\ \ \ H\ \ \ H\ \ \ H\ \ \ H
\end{array}
\qquad \text{caproic acid}
$$

The one exception to the statement of an even number of carbon atoms is isovaleric acid which has five carbon atoms in its chain.

Fats may be either saturated or unsaturated and with or without branches.

When a fat is formed from the esterification of a glycerol and fatty acid, the general structure is:

$$
\begin{array}{c}
\ \ \ \ \ \ \ \ \ \ \ \ \ \ \ O\ \ \ \ \ \ \ \ \ H \\
\ \ \ \ \ \ \ \ \ \ \ \ \ \ \ ||\ \ \ \ \ \ \ \ \ | \\
\text{radical (1)} - C - O - C - H \\
\ | \\
\ \ \ \ \ \ \ \ \ \ \ \ \ \ \ O\ \ \ \ \ \ \ \ \ | \\
\ \ \ \ \ \ \ \ \ \ \ \ \ \ \ ||\ \ \ \ \ \ \ \ \ | \\
\text{radical (2)} - C - O - C - H \\
\ | \\
\ \ \ \ \ \ \ \ \ \ \ \ \ \ \ O\ \ \ \ \ \ \ \ \ | \\
\ \ \ \ \ \ \ \ \ \ \ \ \ \ \ ||\ \ \ \ \ \ \ \ \ | \\
\text{radical (3)} - C - O - C - H \\
\ | \\
\ H
\end{array}
$$

The radicals may be the same or different groups but correspond to fatty acid chains with the last carboxyl group removed.

Common soap is formed by reacting a fat with sodium hydroxide (NaOH) so that the resulting molecule is $C_{17}H_{35}COONa$.

Synthetic detergents are similar except the fat is replaced by long chain hydrocarbon ($C_{17}H_{35}O-$) which has been reacted with a sulfur trioxide group ($-SO_3$) and sodium (Na). The resultant molecule is $C_{17}H_{35}OSO_3Na$.

Natural waxes are different from natural fats in that complex long chain alcohols replace the glycerol component of the molecule. Household "waxes" such as paraffin or carnauba used in floor and auto polishes are really not chemical waxes at all. They are esters.

$$\overset{\displaystyle O}{\overset{\displaystyle \|}{C_{25}H_{52}C}} - O - C_{26}H_{54} \qquad \text{beeswax}$$

$$\overset{\displaystyle O}{\overset{\displaystyle \|}{C_{27}H_{54}C}} - O - C_{26}H_{54} \qquad \text{carnauba}$$

ALDEHYDES AND KETONES

Aldehydes and ketones are related in that both have a double-bonded oxygen atom attached to one of the carbon atoms in the chain. The aldehydes retain one hydrogen on that same carbon while the ketones have other carbon atom(s) bonded to the carbon carrying the oxygen.

$$\begin{array}{ccc} H & O \\ | & \| \\ H - C - C - H \\ | \\ H \end{array} \qquad \text{acetaldehyde} \qquad \begin{array}{ccccc} H & O & H \\ | & \| & | \\ H - C - C - C - H \\ | & & | \\ H & & H \end{array} \qquad \text{acetone}$$

The common mark of both families, the $-\overset{\displaystyle O}{\overset{\displaystyle \|}{C}}-$, is known as a carbonyl group.

In each of the series, compounds up to chain lengths of C_5 are decreasingly soluble in water. Above that chain length they are practically insoluble. The boiling points of the aldehydes and ketones are higher than the basic paraffins from which they are derived.

The name aldehyde is compounded from the chemical reaction from which they are formed, the dehydrogenation of alcohols. When put together, the *al* from alcohol and *dehyd* from dehydrogenation becomes aldehyde.

Aldehydes are usually liquids with a marked odor. They are highly reactive materials that will attack body tissues. Because of this trait they all irritate the eyes and lungs. In this action they follow their solubility tendency; the lower carbon chain aldehydes tend to attack the eyes and exposed respiratory organs such as the nose and throat, while the higher carbon chain aldehydes, less soluble or insoluble in water, will pass untouched into the lungs where they do their damage.

Ketones are a widely variant group, some toxic, some not, some irritating, others not.

The highly volatile ketones enter the body by way of the respiratory tract to do their damage while the less volatile ketones can do their damage to the body through skin absorption.

Many of the ketones are excellent solvents but are highly flammable. They are, therefore, dangerous.

Like their relations, the aldehydes, the lower chain ketones (up to C_4) are soluble in water while from C_5 up they range from slightly soluble to insoluble.

AMINES

The amines are a highly flammable class of organic compounds that are related to ammonia in that one (or more) of the hydrogen atoms of ammonia are replaced by a hydrocarbon group. If only one hydrogen is replaced it is called a primary amine:

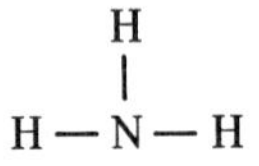

ammonia

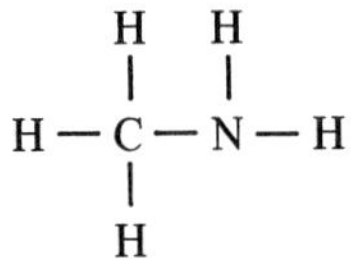

primary amine

If two hydrogens are replaced it becomes known as a secondary amine:

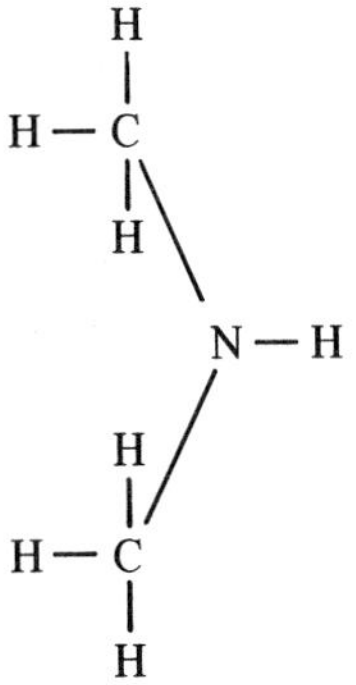

When all three hydrogens on the nitrogen are replaced the compound is known as a tertiary amine:

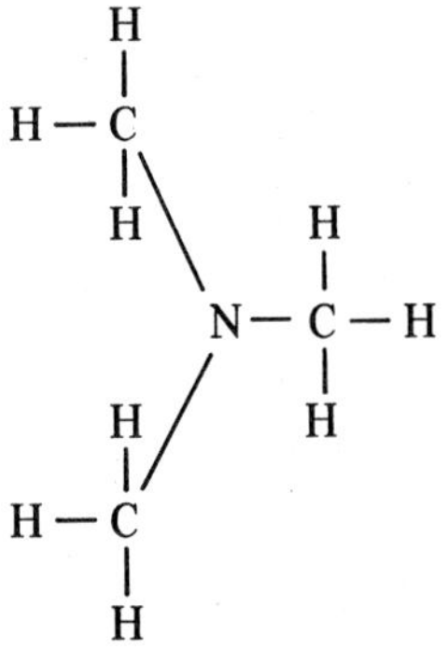

The replacement groups do not have to be similar. Each group may be different.

The smaller chain amines such as methylamine, ethylamine and propylamine have very low boiling points. All of the common amines are toxic and irritating to the skin. Most of them produce vapors with the smell of ammonia.

Aromatic amines are fairly common, existing as primary, secondary and tertiary amines with benzene rings replacing the straight chains of the aliphatic amines:

primary

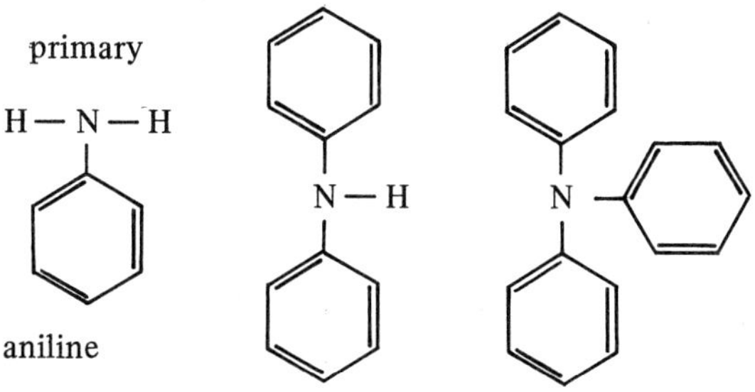

aniline

secondary

tertiary

diphenylamine   triphenylamine

These amines are extremely sensitive to oxidation and, under improper storage conditions, will often decompose merely on contact with air. The aromatic amines may be either liquids or solids.

Hybrid amines also exist which are combinations of aromatic and alphatic amines:

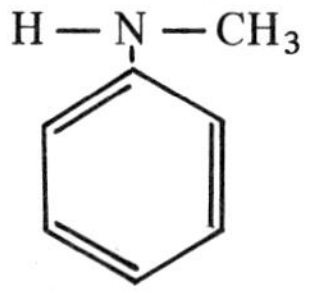

$H - N - CH_3$

dialkylamine

TERPENES

This class of chemical compounds is the basis for two hazardous materials, one liquid, one solid. They have a ring structure but are *not* aromatic—there is no resonance. The liquid is pinene with the following structure:

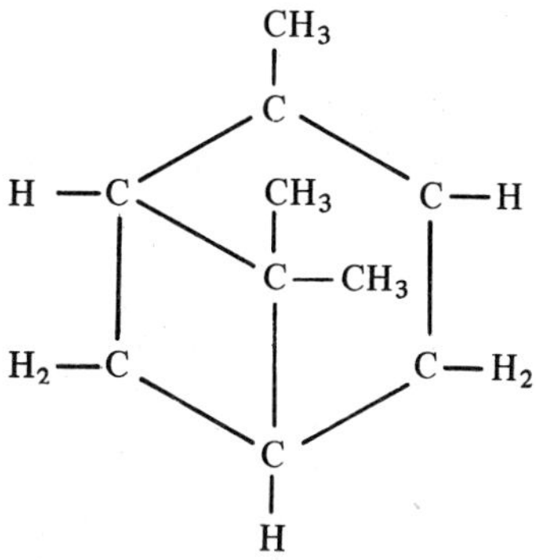

The solid is camphor with this structure:

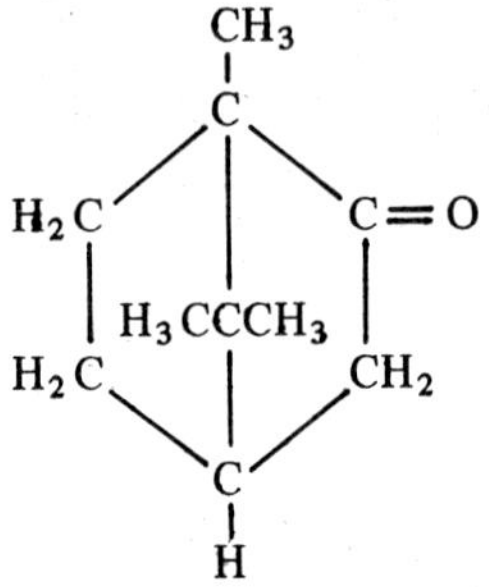

$$
\begin{array}{c}
CH_3 \\
| \\
C \\
\diagup \quad | \quad \diagdown \\
H_2C \qquad\qquad C=O \\
| \quad H_3CCCH_3 \quad | \\
H_2C \qquad\qquad CH_2 \\
\diagdown \quad | \quad \diagup \\
C \\
| \\
H
\end{array}
$$

A closer examination of the camphor molecule shows it has the mark of a ketone, the ketone linkage

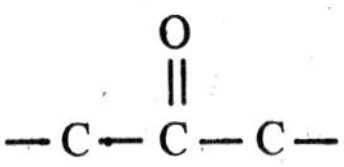

$$
\begin{array}{c}
O \\
\parallel \\
-C-C-C-
\end{array}
$$

It is therefore sometimes classed as a ketone.

PEROXIDES

In addition to carbon, oxygen atoms can also link to other oxygen atoms to form a *very* hazardous class of compounds. The linkage of oxygen to oxygen is relatively rare when compared to the carbon-to-carbon numbers. For this all fire personnel should be grateful. The oxygen to oxygen linkage is a very unstable bond and therefore produces powerful oxidizing agents which are sensitive to impact, shock, and heat.

The oxygen to oxygen linkage produces compounds which are called **peroxides**. Unfortunately, with all too many peroxides, the oxygen to oxygen bond can develop by simple exposure to air over a period of time.

The fundamental peroxide linkage is $-C-O-O-C-$. When this bond is broken there are two fragments produced ($-C-O-$) both with free bonds ready to oxidize just about any other compound.

Organic peroxides are therefore very hazardous materials capable of spontaneous chemical reactions that can lead to disasters. Their high reactivity makes them dangerous not only in fire situations but also to the human body. Most of this class of compounds will irritate the eyes, nose, throat, and lungs. For these reasons they are often dissolved in water or some other solvent to reduce their concentrations for the sake of safety. Even water solutions, however, can be directly hazardous upon skin contact.

According to government regulations peroxides must carry a special DOT (Department of Transportation) label when being transported. The label, a yellow background with a black flame silhouette and black letters which read organic peroxide, guarantees an explosion in an enveloping fire.

Peroxides must not be dropped or shocked in any way. They must be kept cool and out of direct sunlight, which with some peroxides can produce an explosive reaction.

POLYMERS

When discrete molecules link together to form long chains the substance is said to be a **polymer**. The word polymer is derived from the Greek words *poly-* meaning "many" and *-mer* meaning "part." The word polymer therefore distinguishes a multipart, linked molecule from one of its components, a **monomer**. Again, this word is from the Greek word *mono-* meaning "one" or "single" and -mer, as above. To the polymer chemist the word "mer" usually is used to express the simplest chemical molecule by which a polymer may be expressed.

The degree of polymerization is defined as the number of repeat molecules that go to make up the polymer. The molecular weight of the polymer then is simply the molecular weight of the repeated molecule times the degree of polymerization. For example, polystyrene is formed from repeated styrene molecules:

And polystyrene is

"R" indicates a repeat any number of times.

The molecular weight of styrene is:

$$8 \text{ carbons} \times \text{atomic weight of carbon} = 8 \times 12 = 96$$

$$8 \text{ hydrogens} \times \text{atomic weight of hydrogen} = 8 \times 1 = 8$$

The total is 104, the weight of one mer of styrene.

If the basic styrene molecule is linked 2000 times (the degree of polymerization) then the molecular weight of that polystyrene would be 104 × 2000 or 208,000.

The linkage of the polymer chains may be one, or a combination of two or more, of three basic types of linkages (Figure 3.4). These polymers are three dimensional, kinked, coiled, folded back, and twisted in every manner.

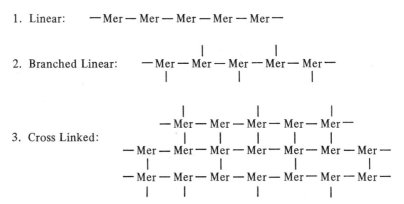

Figure 3.4. The Basic Types of Polymer Chains.

Polymers may be synthetic or natural. Cellulose is an example of a linear polymer occurring in nature, while nylon is a good example of a synthetic or man-made polymer, created for a specific job.

Tailoring of high molecular weight polymers is possible because each of the three polymer types exhibits fundamentally different properties that are attributable not only to the component elements but to the type and size of polymer that is formed. Organic chemists have learned how to replace one element of a molecule with another, to make the type of linkages occur where they want them, and to control the average degree of polymerization. Thus a polymer can be made for specific uses or jobs.

In addition to being able to polymerize molecules of the same type, it is also possible to link unlike molecules and achieve special properties. For example, both styrene and butadiene can be polymerized, to form polystyrene and polybutadiene. They can also be made to react together to form a mixed polymer, butadiene-styrene copolymer:

$$\begin{array}{c}
\text{H} \quad \text{H} \quad \text{H} \quad \text{H} \quad \text{H} \quad \text{H} \quad \text{H} \quad \text{H} \quad \text{H} \quad \text{H} \quad \text{H} \quad \text{H} \\
| \quad | \quad | \quad | \quad | \quad | \quad | \quad | \quad | \quad | \quad | \quad | \\
\text{R} - \text{C} - \text{C} = \text{C} - \text{C} - \text{C} - \text{C} - \text{C} - \text{C} = \text{C} - \text{C} - \text{C} - \text{C} - \text{R} \\
| \qquad\qquad\quad | \quad | \qquad\quad | \qquad\qquad\quad\ | \quad | \\
\text{H} \qquad\qquad\quad \text{H} \quad \text{H} \qquad \text{H} \qquad\qquad\quad \text{H} \quad \text{H}
\end{array}$$

This molecule is better known as Buna S rubber.

Although Buna S has been used for automobile tires it was found to have very poor resistance to high heat as is generated by heavy trucks. The chemists then tailored a new chloroprene molecule into a polymer–neoprene:

$$\begin{array}{c}
\qquad\quad \text{Cl} \qquad\qquad\qquad\ \text{Cl} \\
\qquad\quad | \qquad\qquad\qquad\ | \\
\text{H} \qquad\ \text{H} \quad \text{H} \quad \text{H} \qquad\ \text{H} \quad \text{H} \\
| \qquad\ | \quad | \quad | \qquad\ | \quad | \\
\text{R} - \text{C} - \text{C} = \text{C} - \text{C} - \text{C} - \text{C} = \text{C} - \text{C} - \text{R} \\
| \qquad\qquad\quad | \quad | \qquad\qquad\quad | \\
\text{H} \qquad\qquad\quad \text{H} \quad \text{H} \qquad\qquad\quad \text{H}
\end{array}$$

PLASTICS

The oldest of the man-made molecules goes back to about 1870 when a man named Hyatt was looking for a substitute for ivory which was becoming scarce and therefore expensive. At the same time the game of billiards was growing so popular that it was becoming impossible to supply the demand for billiard balls from available elephant tusks. Hyatt worked with the natural polymer cellulose that exists in wood, cotton, and so many other natural substances. Wood averages about 50 percent cellulose depending upon the type of wood, while cotton is about 90 percent cellulose.

Hyatt managed to replace each of the three hydroxyl ($-$OH) groups on the basic molecule with a nitrate group ($-$ONO$_2$) by causing the cellulose to react with nitric acid.

Under normal conditions, without special chemical inducement, the polymer cellulose can be made to contain 10 percent to 12.5 percent nitrogen. If chemically forced, the reaction will yield cellulose that contains a maximum of 13.5 percent nitrogen. Hyatt's nitrated cellulose therefore contained between 10 and 12.5 percent nitrogen. He found it was soluble in alcohol. However when he mixed his nitrated cellulose with the natural oil of the camphor tree he suddenly had a plastic mass that was pliable but yet seemed to hold its shape well. That mass was the first celluloid. It was naturally white, as is ivory, and with further treatment it solved Hyatt's problem.

Cellulose nitrate still has a great many uses in our modern world. When dissolved in alcohol it is used as a coating for pliable materials such as cloth to form high gloss finishes. It is also the prime component of modern smokeless propellants for guns, large and small, and has been used for photographic film and nail polishes.

All cellulose nitrates are highly flammable and, under some conditions, explosive. However, when cellulose reacts with acetic acid containing no water (acetic anhydride), cellulose acetate is formed. Cellulose acetate has most of the desirable properties of the nitrate but is not nearly as flammable and is not explosive. It is therefore used for most modern film and other nonflammable products.

ACRYLICS

Some compounds polymerize to an extremely high degree. For example methyl methacrylate has the structure:

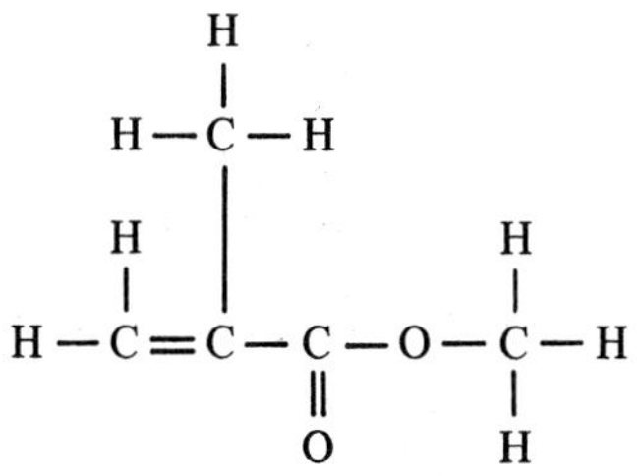

Under the influence of heat and the proper catalyst the polymerization process seems to go on almost indefinitely forming a linear molecule with many branches by fracture of the $-C=C-$ double bond and combining to form methyl methacrylate polymers. These polymers in their molded forms include Lucite and Plexiglas.

VINYL POLYMERS

Vinyl chloride polymerizes in the same manner that methyl methacrylate does. Vinyl chloride has the structure

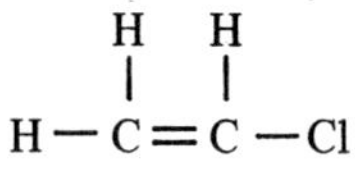

It polymerizes by the fracture of the same double bond. The degree of polymerization is almost infinite.

Vinyl polymers are the basis for a great many decorative plastics. In an enveloping fire they will decompose to form toxic gases and are therefore hazardous.

ORGANIC REACTIONS

The chemical reactions which hydrocarbon molecules undergo can be divided into two groups, **ionic** and **free radical.**

The first type, ionic, is identical to the inorganic reactions discussed earlier with the formation of charged particles called ions. In the case of inorganic ions they are usually single elements. In organic reactions, however, the ions may be whole molecules with a single excess charge. In both inorganic and organic ionic reactions the covalent bond breaks to form a cation (positively charged particle) and an anion (negatively charged particle).

In free radical type reactions the covalent bond breaks in such a way as to provide fragments, each with an excess electron. These negatively charged fragments are called free radicals. Quite often free radical generation can exceed the free radical removal (the formation of other compounds) and the reaction can run away, perhaps resulting in an explosion. When the free radical removal exceeds the generation, the reaction is slowed down, or may even be brought to a halt. This is the mechanism of fire extinguishment. It will be discussed in greater detail later.

SIZE, CHARACTERISTICS, AND BOILING POINT

A study of molecular size indicates that there is an important relationship between the number of carbon atoms, the groups that are attached and that very important characteristic for fire personnel, boiling point.

Figure 3.5 is a graph of carbon chain lengths for the first seven saturated chain hydrocarbons. Unsaturated bonds have a slight influence on the boiling point of a given chain length but do not change it radically. However, when a hydrogen atom is replaced with a hydroxyl group (—OH), the boiling point is altered considerably, as can be seen in the shift of the curve from the alkane to the equivalent alcohol (Figure 3.6).

The ether linkage (—C—O—C—) reduces the boiling point from that of the alcohol but at least up to diethyl ether the boiling point is less than the parent chain. For diethyl ether and higher chain lengths the boiling point is close to the straight chain, unaltered hydrocarbon.

The ketone linkage

$$\overset{\textstyle O}{\underset{\textstyle -C-}{\|}}$$

elevates the boiling point but to nowhere near the extent of the alcohol linkage.

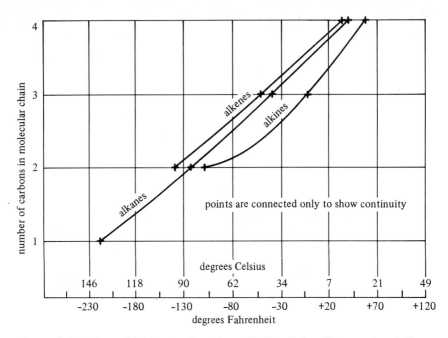

Figure 3.5. Effect of Molecular Length on Boiling Point (Saturated and Unsaturated Straight Chains).

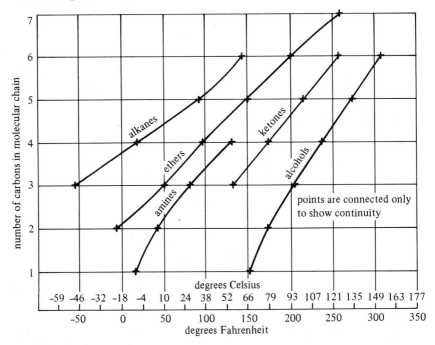

Figure 3.6. Effect of Molecular Length on Boiling Point (Short, Aliphatic Hydrocarbon Families).

REVIEW QUESTIONS

1. What is the scope of organic chemistry?
2. What is a Kekulé diagram? Give an example.
3. What do the suffixes -ane, -ene, and -ine indicate when used in the name of an organic chemical?
4. What is an isomer? How is it related to others of the same family?
5. What is a radical in organic chemistry? Draw a methyl radical, an ethyl radical, and a propyl radical.
6. List the order used when naming an organic molecule.
7. What is the difference between aliphatic and aromatic compounds?
8. Explain the term resonant hybrid.
9. Give the characteristic linkages that differentiate between alcohols, aldehydes, amines, esters, ethers, ketones, organic peroxides, and organic acids.
10. How is an alcohol related to water?
11. What is a glycol? How does its structure affect its boiling point?
12. What is a glycerol? How is it related to alcohol and glycol?
13. What is an ether? What characteristic does its vapor possess that is important in fighting ether fires?
14. What is an amine? What is the difference between primary, secondary, and tertiary amines?
15. What are peroxides? Why are they so hazardous?
16. What is a polymer? What is a mer?
17. What are the basic types of polymers?
18. Detail the difference between ionic and free radical reactions.
19. Discuss the relationships between the sizes and boiling points of organic compounds.

Oxidation/ 4

Fire has been defined as a rapid oxidation process or the combining of a substance with oxygen in the atmosphere. This is basically, but not universally, true. While our atmosphere is composed of about 79 percent nitrogen, 20 percent oxygen, and 1 percent rare bases, carbon dioxide, and water, the oxygen for fire can come from other sources besides the atmosphere. Oxygen is essential to our existence, but it is also the one element that is common to most fires and therefore can be a hazard.

AGENTS

It was explained earlier that chemists designate the loss of electrons in a chemical reaction as oxidation and the gain of electrons as reduction. Following this procedure, any substance that *causes* the loss of electrons is called an **oxidizing agent.** Just as an insurance agent does not, himself, insure anyone, but instead promotes or sees to it that someone else is insured by a company he represents, so an oxidizing agent promotes the oxidation, or loss of electrons, in another substance. In the oxidation reaction, however, in order to cause the loss of electrons, that agent must itself gain electrons. So also any substance that promotes the *gain* of electrons is called a **reducing agent**, and in order to cause the gain, it must lose electrons.

process	action
oxidation	loss
reduction	gain

agent	process
'oxidizing	gain
reducing	loss

The classification of a substance as an oxidizing agent is made purely on its ability to take up electrons. The rating of this ability to carry out this action is based upon the energy change of the process wherein the electrons are absorbed. There is always an excess of energy in the process of oxidation so the reaction is always exothermic. Ozone is an excellent oxidizing agent because it has a large energy content. Much of this energy is released in its oxidation reactions.

In order to be able to classify oxidizing agents, the chemist details where the electrons go in an oxidation-reduction reaction. In this technique, the chemist assigns a number to each elemental atom and calls it the **oxidation number**, also known as the **oxidation state**. This number is defined as the charge which seems to be associated with each atom. The word "seems" is used because of the sharing we have discussed previously. The number is assigned according to some very arbitrary rules:

- Electrons shared by two dissimilar atoms are assigned to the more electronegative atom.
- Electrons shared by two similar atoms are divided equally between the sharing atoms.

As an example, hydrogen never exists as a single atom but always as two atoms linked together in a molecule (H_2). The oxidation number of hydrogen then, according to Rule 2 above, is zero: one electron is counted with each atom. Because the hydrogen core has a positive charge of +1 with one electron in the outer shell to balance this in each atom, we have an equal number of positive and negative charges in the molecule as well. The resultant charge is zero.

Water is quite different, however. In the H_2O molecule, oxygen is more electronegative (2:1) and therefore, according to Rule 1 above, the shared electrons are assigned to the oxygen. Here the hydrogen atom would have a single positive charge resulting in an oxidation number of +1. The oxygen would then have an oxidation number of −2 when in a water molecule.

Some operating rules can be derived from these observations:

- In most compounds of oxygen, oxygen has an oxidation number of −2.
- In most hydrogen containing compounds, the oxidation number of hydrogen is +1.
- In free elements, the oxidation number of each atom is zero.
- In simple (single atom) ions, the oxidation number is equal to the charge on the ion, positive or negative.

Remembering that an oxidizing agent is one that causes a loss of electrons by gaining electrons itself, and that the efficiency of an oxidizing agent is directly related to the energy change of the electron absorbing process, we can, by means of electrochemistry, rate oxidizing agents according to their efficiency.

This is accomplished by measuring the electrode potentials of the various elements in chemical reactions using standard solutions in a device called a **hydrogen half cell**. Recall that hydrogen's oxidation number is zero, so the potential developed by this half cell is standardized at zero, for reference with other substances.

Some oxidation numbers are

Element	Oxidation Number
aluminum	+3
zinc	+2
lead	+2
magnesium	+2
silver	+1
sodium	+1
chlorine	−1
sulfur	−2
oxygen	−2
phosphorus	−3
nitrogen	−3

Following the reasoning that oxidation is the loss of electrons, we can judge that the more negative the oxidation number, the stronger the oxidizer.

Oxidation-reduction reactions go on around and in us continually. They vary in their individual rates of reaction but nevertheless they continue, some relatively slow, some very fast. They are common because the element oxygen is so plentiful on our planet, and because most chemical reactions involve a gain and loss of electrons. They are so common to the chemist that they have been given a nickname—they are called **redox** reactions.

OXYGEN

The element oxygen is a colorless, odorless gas at normal temperature and pressure. Although it is the most plentiful of all elements on earth, it was not until the late 1700s that it was isolated and recognized. Long before that, however, scientists knew there was something that was common to fire as well as to many reactions in the laboratory. It is generally believed that an English clergyman, Joseph Priestley, whose hobby was scientific investigations, was the first to discover oxygen, identify it, and publish his findings.

Because he was fascinated with the investigation of gases, Priestley developed special apparatus for collecting the gas given off by just about any material that he could get his hands on. His apparatus looked like that seen in Figure 4.1.

When the substance under study gave off gas when heated, the gas would travel through the tube to the collecting jar where it would displace the water.

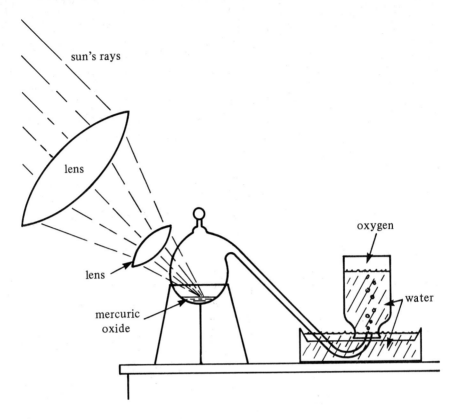

Figure 4.1. Priestley's Apparatus.

When almost all of the water had been forced out of the jar, Priestley would cork the jar and the gas was trapped.

Because he lacked the modern convenience of a laboratory gas burner to apply heat in his experiments, Priestley used a large and powerful magnifying glass to collect the sun's rays and so bring the substances under study to a high temperature.

In August of 1774, Priestley put some material known to the chemists of that time as mercurius calcinatus (to us, mercuric oxide) in his apparatus and focused the sun's rays on it. A gas was given off, displaced the water, and was trapped in the jar. In one of his experiments with the newly collected gas he opened the jar, mouth down, and put a burning candle up into the gas. Instead of the candle snuffing out as it had so often with other gases, the candle this time flared up even brighter. Priestley called the new gas dephlogisted air because he believed he had extracted the "burning causer" from air.

Some time later a French chemist. Anton Lavoisier, became fascinated with Priestley's experiments with dephlogisted air and repeated many of them.

Lavoisier not only confirmed Priestley's results but was able to show that when a substance is burned in air in a closed container, there is a decrease in the total volume of air present. By measuring the amount of air before and after a burning experiment, Lavoisier was able to show that about 20 percent of the contained air disappeared. He reasoned, correctly, that when substances burn in air a new substance is formed which must contain most of the air that disappeared. Thus the modern theory of combustion was born. Lavoisier called Priestley's gas **oxygen**, meaning "acid producer," because he believed this gas was the necessary component of all acids. We know now that it is not.

It has since been shown that oxygen constitutes about one half of all the material on this planet that we can put to analysis. Uncombined it constitutes about 20 percent of our atmosphere, as shown by Lavoisier. In the most plentiful compound on earth, water, oxygen is about 90 percent by weight. Oxygen is also found in animals (60 percent by weight) and in most rocks of the earth's crust. The solubility of oxygen in water, however, is surprisingly low, only about 5 percent. Other physical properties are: boiling point, $-297\,^{\circ}$F ($183\,^{\circ}$C); freezing point, $-360\,^{\circ}$F ($218\,^{\circ}$C); and density, 1.43 grams per liter at $0\,^{\circ}$C.

SOME CHEMICAL PROPERTIES

Fortunately for mankind atmospheric oxygen is not very reactive at normal temperatures. With our uses of wood for our homes, gasoline for our automobiles, solvents in paints, and all the other very common uses of combustible materials, if oxygen were highly reactive there would be no world, at least as we know it today. Oxygen develops a different character at elevated temperatures, however, and will react with just about every other known element except the inert gases, a few of the really inactive metals, and the halogen family consisting of chlorine, bromine, iodine, fluorine, and astatine. Even carbon, which is, as an inorganic chemical, considered relatively inactive, will react with oxygen if the temperature is high enough. All reactions with oxygen are exothermic and therefore liberate heat. When the reaction is fast enough that heat will be accompanied by light. Most of the reactions with oxygen must be started by some sort of energy input which raises the temperature and starts the reaction so that energy is given off. The reaction then becomes self-sustaining.

The common oxidation processes such as combustion, respiration, rusting, and decay are familiar enough to us. They are all exothermic. In some, easily measured heat is produced, in others it is not. This is not to say identical reactions do not produce the same amount of heat when one is slow and the other fast. The difference is, because of the time element, the generated heat is dissipated and the temperature does not rise an easily measurable amount. In all cases involving the same reactants, the amount of heat evolved is identical as long as the chemical process is identical.

OXIDATION RATE

Wood decay and wood burning are identical chemical reactions that produce carbon dioxide (CO_2) and water (H_2O). They are both exothermic, releasing the same amount of heat. The difference is, of course, the rate at which the oxidation process proceeds. In the case of decay it proceeds so slowly that it needs constant starting by microorganisms. The temperature of the material (unless enclosed) never rises to a point where the redox reaction becomes self-sustaining. In the case of wood burning, ignition must be provided by an outside source, but where so many wood molecules are all ignited simultaneously, and are in such close proximity to each other, the temperature of the material rises (from the exothermic reaction) above the ignition point of the rest of the nearby material and the system becomes self-sustaining, even runaway. It will cease only when all material has been oxidized or the system is influenced by external, heat-removing or oxygen-removing, procedures.

The true difference between generating a temperature sufficient for ignition very quickly and very slowly is the **rate of reaction**. Some familiar examples of such rates are:

Process	Rate of Oxidation
iron rusting	slow
wood decay	slow
moist hay decaying, confined	slow-accelerating
gunpowder, unconfined	medium
gunpowder, confined	fast
gasoline vapors, unconfined	fast
gasoline vapors, confined	very fast—explosive
nitroglycerine	very fast—explosive

In most instances the rate of the redox reaction is influenced by its confinement. There are two forms of confinement, temperature confinement or insulation, and pressure confinement. They are related and interdependent.

Temperature confinement, or insulation, is merely a condition in which the heat, generated by an exothermic reaction, is not dissipated. When it does not leave the reaction area, it promotes an increase in the reaction rate allowing the total heat in the system to build up until the reaction runs away. In all chemical reactions, the rate of reaction doubles for every ten-degree increase in temperature.

The second type of confinement, that of pressure, occurs when the redox reaction-generated gases are not allowed to escape. The same upward spiral of pressure begins and accelerates just as it did previously with the temperature. The increase in pressure promotes a faster reaction as well as an increase in temperature, which in turn generates more pressure, accelerating the reaction rate still more. Many substances that merely burn when unconfined will explode when confined and ignited. Modern smokeless gunpowder is an excellent example. One volume of smokeless powder will generate up to 14,000 volumes of gas as it burns. If unconfined it burns at a medium speed. If confined as it is in a cartridge case within the chamber of a rifle, it will generate as much as 50,000 to 60,000 pounds per square inch of pressure in 0.0005 seconds when ignited.

Another important factor influencing the rate of oxidation is the particle size of the material being oxidized. No knowledgeable person would ever try to start a fire with three-inch diameter logs, knowing that they would never catch. However, that same three-inch log can be whittled into a few shavings and those used to start slightly larger branches until there is enough heat in the fire to bring larger and larger logs up to their ignition point. This is merely an example of the influence of particle size on the rate of reaction. The smaller the particles, the smaller the mass that must be raised to the ignition point, and the less heat input required to do the job. The more molecules exposed to the initiating energy, the easier it is for the loss or gain of electrons to begin and the redox reaction to proceed. This is called **surface area** influence. Gases react more easily than solids, because in a gas, every molecule is in a sense exposed, and is free and available to react. The surface area of a gas, which is actually the exposed outer-shell electrons of every molecule, approaches infinity.

OXIDES

When elements unite with oxygen to form simple inorganic compounds, the resultant substances are called **oxides**. The reaction of many organic compounds with oxygen will also produce oxides if the reaction proceeds far enough. The burning of wood is actually the oxidation of the carbon atoms in the complex compound which we call wood. When this compound is oxidized it produces carbon dioxide and carbon monoxide as well as water. Carbon dioxide is also formed by the slow oxidation of cells in the lungs, and then exhaled. Iron forms an iron oxide, commonly known as rust, when exposed to water and air.

Oxides can be divided into two distinct families depending on whether the combining element is a metal or a nonmetal. When a metal oxide is formed it can react with water to form an **alkaline** solution or, as it is commonly called, a **base**. When nonmetallic oxides react with water they form an **acid**. The metal sodium reacts thus:

$$4Na + O_2 \longrightarrow 2Na_2O \qquad \text{(sodium oxide)}$$

$$2Na_2O + 2H_2O \longrightarrow 4NaOH \qquad \text{(sodium hydroxide) \quad base}$$

The metal aluminum reacts:

$$4Al + 3O_2 \longrightarrow 2Al_2O_3 \qquad \text{(aluminum oxide)}$$

$$Al_2O_3 + 3H_2O \longrightarrow 2Al(OH)_3 \qquad \text{(aluminum hydroxide) \quad base}$$

The nonmetal sulfur reacts:

$$S + O_2 \longrightarrow SO_2 \qquad \text{(sulfur dioxide)}$$

$$SO_2 + 2H_2O \longrightarrow H_2SO_3 \qquad \text{(sulfurous acid)} \quad \text{acid}$$

The story is told, that a chemical plant had a major problem just because of the above reaction. In one of its manufacturing processes sulfur dioxide (SO_2) was piped into a waste stack and vented into the atmosphere. Not long after the process was started, inhabitants in the town found that the family wash hanging on the lines would sometimes rot even though almost new. When the clothes did not shred while being taken off the line, they would become so weakened that the mere act of putting them on would cause them to fall apart. Because of the noxious odors that frequently came from the chemical plant it was blamed for the rotting of the clothes. The chemists found the destruction always occurred on very moist days. The sulfur dioxide from the vent stack was reacting with the water vapor in the air to form sulfurous acid, which, in turn, oxidized the clothes.

example: Another nonmetallic oxide is formed:

$$C + O_2 \longrightarrow CO_2 \qquad \text{carbon dioxide}$$

$$CO_2 + H_2O \longrightarrow H_2CO_3 \qquad \text{carbonic acid} \quad \text{(acid)}$$

It should be noted from the above examples that all bases have the hydroxyl (OH) group while acids have hydrogen in their molecules. These are the marks of inorganic acids and bases. The strength of an acid is due to the amount of hydrogen ion (H^+), or of a base, the hydroxyl ion (OH^-), that is present in the solution. The measurement of these quantities is the pH (power of hydrogen) of the solution. The more hydrogen ion available, the stronger the acid, while the less hydrogen ion and the more hydroxyl (OH) ion available, the more alkaline or basic is the solution. When equal amounts of H^+ and OH^- are in the solution and they are the only groups present, the solution is neutral and we call it pure water.

Because the range of values is so extensive, the measure of pH is put on a logarithmic basis. Because the actual ratio of H^+ to OH^- ions in a standard solution may be very small (such as 0.0000001), the expression for that quantity has a negative exponent. Pure water, with a hydrogen ion concentration of 10^{-7} per standard quantity, has a pH of 7. Because of the negative exponent, as the hydrogen ion concentration increases, the exponent gets smaller. A solution with a hydrogen ion concentration of 10^{-3} would therefore have a pH of 3. Conversely, when the hydroxyl ions outnumber the hydrogen ions, so that the hydrogen ion concentration is less than 10^{-7} per standard quantity, for example 10^{-12}, the solution is considered alkaline (basic) with a pH of 12.

Graphically pH might be presented as follows:

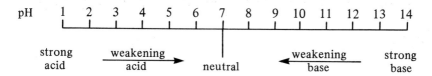

OXIDIZERS

While most of the fires with which we are familiar use the oxygen in the atmosphere as their oxidizer, there are many fires that obtain their supply of oxygen or their oxidizing agent from chemical compounds. These compounds, rich in oxygen or containing a strong oxidizing agent other than oxygen, need no air whatever for their redox reactions.

Smokeless gunpowder and both solid and liquid propellant rocket motors are examples of formulations that provide their own oxidizing agents.

Most of the strong oxidizers are stable at room temperatures and need to reach a relatively high temperature before they begin to break down and provide the oxygen, or they need a strong reducing agent with which to react. There are, however, many that are unstable even at room temperature.

OZONE

One of the most common oxidizers, and one that is quite unstable, is a variant form of oxygen called ozone. This form of oxygen has different physical properties and even a different energy content. Because it is pure oxygen and yet differs from the parent element, it is called an **allotropic form**. Many elements exhibit this phenomenon.

Ozone was discovered in 1785 by a chemist named Van Marum but wasn't really understood until almost 75 years later by a man named Schönbein. It was the latter who gave it its present name ozone, from a Greek word meaning "to smell." Experiments by Andrews in 1856 proved it was pure oxygen and paved the way for Soret to show, in 1866, that the formula was O_3.

Ozone is generated in the atmosphere by lightning flashes or by other spark-type electrical discharges. The quantities produced, however, are very small and, because ozone is such an excellent oxidizer, react almost immediately with water vapor to form hydrogen peroxide (H_2O_2).

Ozone is a bluish colored gas with an intense, unmistakable, odor. Research has shown that as little as one part of ozone in ten million parts of air can be detected without any special apparatus save one's nose.

One of the most indicative properties of ozone is the energy that is absorbed into each molecular weight of ozone that is formed. That energy is about 35 kcal for every 48 grams. It is this potential energy, stored in the molecule, that makes it react so readily. The release of this much energy by every 48 grams produces reactions that occur very fast.

While some substances, such as natural rubber, will oxidize over a long period of time and so deteriorate, exposure to ozone will produce the same results in a very short period.

Ozone's high reactivity makes it extremely useful in removing odors of all kinds, killing bacteria in water supplies, as an excellent bleaching agent, and for many other industrial uses.

NITRATES

All reactions between nitric acid (HNO_3) and alkaline compounds form salts, which are universally powerful oxidizing agents.

Because they are so common and because, as strong oxidizers, they contain considerable energy, they must be considered hazardous. While it may seem extreme to brand a common, inorganic fertilizer as hazardous it is nevertheless true. Ammonium nitrate is a very common and very fast-acting fertilizer. It was, however, the explosive of the Texas City, Texas, disaster of 1947. A ship, docked there while unloading, blew up in a series of devastating explosions. When the smoke had cleared, 512 people were dead and over 3000 injured. The property loss was estimated at over $70,000,000.

The dynamite business was all but wiped out when enterprising farmers discovered that ammonium nitrate mixed with crankcase oil is every bit as effective as the commercial product for blasting tree stumps and other common obstacles. The mixture is listed by the United States government Department of Transportation as AN/FO (ammonium nitrate/fuel oil), a hazardous material.[1]

A relative of ammonium nitrate, potassium nitrate, is the oxidizer that makes black gunpowder a propellant. In this era of nostalgia, it is a very popular propellant with the muzzle loading enthusiasts. Although it is hazardous, many of these buffs keep as much as fifty pounds in their homes, enough to destroy an average-size house. According to United States federal regulations, however, that quantity is legal.

Hydrogen peroxide (H_2O_2) is a commonly used and very strong oxidizing agent. It is classed as a peroxide because the molecule contains more oxygen than normal valency dictates it should. In other words, the hydrogen in this compound seems to have a valence of *two* rather than one as it does in its normal compounds. Because of this abnormality, the compound is unstable at normal temperatures and is therefore highly reactive. When it decomposes it releases over 46 kcal of heat, thus:

$$H_2O_2 \longrightarrow 2H_2O + O_2 \uparrow + 46.3 \text{ kcal}$$

[1]U.S. Department of Transportation, *Hazardous Materials Regulations,* Code of Federal Regulations, Vol. 49, parts 100–199.

Because sunlight will cause this decomposition reaction, hydrogen peroxide is usually kept in brown glass bottles. As the formula above indicates, gas is generated in the decomposition. It is therefore a good safety precaution to open bottles of hydrogen peroxide with great care. Many industrial accidents have been reported in which a large container of the peroxide was opened too quickly with a resultant spray of oxidizer onto hands, face, or clothing. Usually the hydrogen peroxide sold for household use contains only three to four percent hydrogen peroxide in water, and is therefore much safer to handle.

While the above oxidizers are inorganic, there are many organic oxidizers far more powerful, which will be discussed in later chapters.

NONOXYGEN OXIDATION

Strong acids such as concentrated sulfuric are very good oxidizing agents and their reactions illustrate chemical oxidation without the presence of elemental oxygen. Even relatively inactive elements such as carbon or copper can be oxidized when placed in hot, concentrated sulfuric acid:

$$C + 2H_2SO_4 \longrightarrow 2SO_2 + CO_2 + 2H_2O$$

Reading this chemical shorthand we find that one atom of carbon reacts with two molecules of sulfuric acid and the reaction produces two molecules of sulfur dioxide, one molecule of carbon dioxide, and two molecules of water.

Copper reacts with hot, concentrated sulfuric acid as follows:

$$Cu + 2H_2SO_4 \longrightarrow SO_2 + 2H_2O + CuSO_4$$

Translating this we find that one atom of copper reacts with two molecules of sulfuric acid to produce one molecule of sulfur dioxide, two molecules of water, and one molecule of copper sulfate.

An organic example of nonoxygen oxidation is that of the reaction of sugar with sulfuric acid:

$$C_{12}H_{22}O_{11} + 11H_2SO_4 \longrightarrow 11H_2SO_4 \cdot H_2O + 12C$$

This reaction formula indicates that eleven molecules of sulfuric acid are required to fully oxidize every molecule of sugar. It also indicates that this oxidation process produces eleven molecules of sulfuric acid with a molecule of water tied to each, and twelve atoms of carbon. The latter product, a mass of black carbon, is the only visible remains of the reaction.

Nitric acid is also an excellent oxidizing agent and as such is often used in liquid-propellant rocket motors with any one of several organic fuels (reducing agents). In contrast to sulfuric acid, which must be hot in order to oxidize copper, cold nitric acid will, even when dilute, oxidize copper, thus:

$$3Cu + 8HNO_3 \longrightarrow 4H_2O + 2NO + 3Cu(NO_3)_2$$

THE HALOGENS AS OXIDIZING AGENTS

Fluorine, chlorine, bromine, and iodine make up a family of compounds called halogens that are nonoxygen oxidizing agents to varying degrees, with fluorine as the best and iodine as the poorest. All four elements, nevertheless, are very irritating to the human mucous membranes, with bromine the most dangerous because it burns the skin as well as the eyes and nose.

The action of these oxidizing elements may be explained by their relative electronegativity and their affinity to gain an electron causing the loss or oxidation of some other material. Fluorine has the greatest ability to pick up an electron and is actually the strongest oxidizing agent known of all the elements.

Chlorine is not as strong an oxidizing agent as fluorine but is, nevertheless, a very hazardous material. It is quite common in industry and health care materials, making it all too often a gas that fire safety personnel must face in both fire and nonfire situations.

Bromine and iodine are weaker oxidizing agents than either fluorine or chlorine but are still very hazardous materials because they react so easily with body fluids to form oxidizing acids which in turn can produce serious burns both internally and externally. Bromine and chlorine, for example, react with water as follows:

$$Br_2 + H_2O \longrightarrow HBr + HBrO \qquad \text{(hypobromous acid)}$$

$$Cl_2 + H_2O \longrightarrow HCl + HClO \qquad \text{(hypochlorous acid)}$$

Fluorine reacts with water to produce hydrofluoric acid:

$$2F_2 + H_2O \longrightarrow 4HF + O_2$$

Any solution of a halogen and water results in strong oxidizing agent action and must be treated as a hazardous material.

Chlorine and fluorine react explosively with hydrogen as follows:

$$H_2 + Cl_2 \longrightarrow 2HCl \qquad \text{(hydrochloric acid)}$$

$$H_2 + F_2 \longrightarrow 2HF \qquad \text{(hydrofluoric acid)}$$

These reactions will not take place in the dark but when the mixed hydrogen and halogen are exposed to some sort of trigger energy such as sunlight, it runs away with explosive violence.

SULFUR

Another very common nonoxygen oxidizing agent is elemental sulfur. Because it has only six outer orbital electrons in the energy level that can hold as many as eight, sulfur, like oxygen itself, usually exhibits an electronegativity of -2. Sulfur is often considered to be part of the oxygen family as are selenium and tellurium. Both selenium and tellurium are also oxidizing agents, but are relatively weak.

Sulfur seems to have a split personality when it comes to gaining or losing electrons. When reacted with metals, it is an oxidizing agent and will gain an electron; with nonmetals, it will more often than not give up an electron. It is, however, considered an oxidizing agent.

SUMMARY

In summary, then, any element or compound which is electronegative and therefore has the ability to pick up an electron must be considered an oxidizing agent whether it contains oxygen or not. The relative ability to pick up the electron determines its strength as an oxidizer.

BIOCHEMICAL OXIDATION

All animals require the input of oxidizable materials in the form of carbohydrates, fats, proteins, and inorganic salts. These materials are oxidized to develop the energy and the necessary chemical compounds that keep the life process intact. Oxidation of fat in the body is by far the best source of energy we have. It has been determined that one ounce of fat develops about 270 kcal of energy when oxidized within the average human. In addition to this energy, carbon dioxide and water are produced. This CO_2 and H_2O are extracted from the system by the lungs, which exchange oxygen for these compounds in the breathing process.

This vital chemical exchange takes place by means of the more than 600,000,000 alveoli or air sacs in the lungs. These sacs are made up of intermeshed networks of very small blood vessels or capillaries. In this network, the blood is cleansed of the products of oxidation, the carbon dioxide and water, and reoxygenated for its trip back to the various parts of the body.

Any chemical interference with the operation of the air sacs, either because of oxygen deficiency in the inhaled air or because of compounds in that air occluding the air sacs, results in a reduction in the oxidation supply to the body and becomes, therefore, a hazard. A simple oxygen reduction of only four to five percent in the air inhaled can be hazardous, depending on the physical condition of the body and the amount of physical work being done. There are, in addition, innumerable compounds, both organic and inorganic, that either attack or coat the air sacs. The more common ones will be discussed later.

Other very important oxidation reactions that take place within an animal body are part of the **metabolic functions**. This is the terminology used to describe the processes wherein digested foods are used either to supply energy (from exothermic reactions) or to be transformed into organic compounds called bone or tissue. Actually a food molecule can be used for both metabolic functions. It might first become tissue, and then at some future time be oxidized.

The oxidation of carbohydrates and fats as part of the metabolic function of providing energy and body heat is not a simple chemical reaction. It is a complicated, multistep procedure occurring at relatively low temperature. It does, however, develop the same amount of energy as those same food compounds would if they were burned in an open flame.

As in the breathing process, any compound or element that blocks or in any way interferes with these internal oxidation processes must be considered hazardous. These too will be discussed later.

REVIEW QUESTIONS

1. Define oxidation.
2. Distinguish between oxidizing agent, reducing agent, oxidation, and reduction.
3. What is an oxidation state?
4. List the rules used to assign oxidation numbers.
5. Give the oxidation numbers for sodium, chlorine, nitrogen, gold and oxygen.
6. What is a redox reaction? Give an example.
7. What property of oxygen allowed Priestley to collect it by displacing water?
8. Define rate of reaction. Give some examples.
9. What conditions influence reaction rate and in what manner?
10. What is surface area influence? Explain.
11. What is an oxide? Give some examples.
12. What is formed by the reaction of a metallic oxide and water? Of a non-metallic oxide and water?
13. What is pH? Explain.
14. Are solutions with the following pH acid or alkaline: 2, 7, 14, 12, 3, 5, 4, 8?
15. Define allotropic form. Explain.
16. Why is ozone a good oxidizing agent?
17. What is a nitrate? How may it be formed?
18. Translate the following:

$$2Na_2O_2 + 2H_2O \longrightarrow 4NaOH + O_2$$

19. What is a peroxide? Explain.
20. Briefly explain biochemical oxidation and its importance to human life.

Combustion/ 5

FIRE

Fire is a visible result of the phenomenon of combustion, and combustion is the result of rapid oxidation. Combustion is therefore a chemical process (oxidation) which evolves energy in visible and invisible forms. The invisible form we call heat; the visible, light.

All combustion is oxidation but not all oxidation is combustion. Combustion is a specific type of oxidation. While oxidation is always exothermic, only when the amount of energy given off reaches levels high enough to produce light as well as heat do we call the process combustion.

Like oxidation, combustion does not always involve combination with oxygen. For example, flame accompanies the decomposition of the rocket fuel hydrazine, but there is no combining with oxygen. Under the proper conditions, chlorine will oxidize hydrogen so rapidly that an impulse of light is emitted by the resultant explosion. There is no oxidation involved in this combustion.

The fire triangle diagram is often mistakenly labeled: one side, fuel; one side, heat; one side, oxygen. The latter side should be labeled oxidizer, rather than oxygen, because oxygen or no, if electrons are lost rapidly enough there will be a fire.

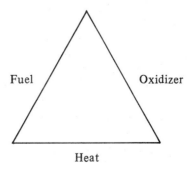

Fuel

Oxidizer

Heat

With the word "Fire," beginning with the letter *F,* the whole idea might be better presented:

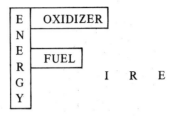

When those three components come together in one place, it can be the beginning of a fire. Note that in the *F* diagram, the first letters of the three elements of fire spell *FOE.* Fire certainly is.

The temperature at which a substance begins to oxidize so rapidly that it produces heat and flame, becoming self-sustaining in its redox reaction, is called the **kindling point** or the **ignition point**.

Some common ignition points are:

Material	Ignition point	
	°F	°C
paper	365	183
cotton	440	227
cellophane	470	243
wood	475	246
methyl alcohol	800	427
cooking gas	1225	663

One consideration just as important as temperature is the particle size or mass. While the particle size does not change the kindling point as far as the input temperature is concerned, it does change the amount of energy (heat) required to start a reaction. The smaller the particle size, the smaller the mass, and the less heat input needed to bring that mass up to its ignition point. A good example of how mass effects the heat required to bring an object to its kindling temperature is that of a cigarette lighter. The lighter flame will heat the point of a pin to red-hot and yet never do any more than warm the point of a crowbar. The masses of the two are quite different. For the same reason, vapors react with much less input than solids. The vapors have particles of molecular size together with an almost infinite number of surface area electrons ready to react. Coal dust, wood dust, flour dust, and even lubricants such as magnesium stearate dust will explode with great violence if they are dispersed in air and exposed to just a small spark or other source of ignition. An explosion in a chewing gum factory on Long Island in November of 1976 killed several people and injured many others when, as reported by the fire department there, the dust from magnesium stearate, being used as a special lubricant in the process, was ignited by a spark from a nearby machine.

This surface area influence is also the reason why fuel in an oil burner is atomized by special nozzles before it is burned. The carburetor of an automobile increases the surface area of the gasoline, as well as mixing it with air, for better combustion. While atomization does not really produce particles of atomic size, it does nevertheless increase the surface area tremendously. A solid like coal, or a liquid such as oil or gasoline, when atomized will burn very much like a gas with its almost infinite surface area.

FLAME

When light emission is made an important part of the definition of combustion, we really define that term, at least in part, in relation to the human eye. Light is merely a small part of the total energy spectrum, that which excites certain components of the human optical system. The human eye is an energy receiver sensitive to a narrow band of energy emissions.

The total energy spectrum could be depicted as in Figure 5.1.

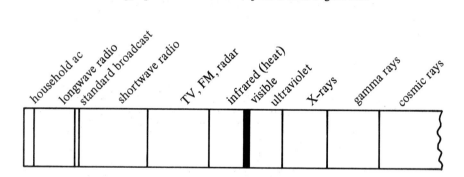

Figure 5.1. The Energy Spectrum.

Light is radiated energy of a band of frequencies that affects sensitive portions of the human body. The energy may be reflected from an object or may be developed within the object, but in either case we see it because radiated energy is involved.

Modern theory explains light as vibrations with each color having its own frequency or number of vibrations per second. The lowest frequency the average human eye can see is dark red light which has a frequency of about 420 million, million cycles per second. The highest frequency we can see is called violet, and has a frequency of about 75×10^{13} cycles per second. Compare this with the Standard Broadcast Band radio which receives radiated energy that has frequencies between 550 and 1600 thousand cycles per second. Citizen Band radios operate at about 27 million cycles per second.

When an oxidation reaction is rapid enough to give off both heat and light, the energy appearing as light is always a very small part of the total evolved energy. It is nevertheless an important indicator. Just as an indicator light on a control panel warns of some condition, so the flame warns that an oxidation reaction is in progress with sufficient energy output to be self-sustaining. This is not to say that every self-sustaining redox reaction will produce visible emissions. Some extremely rapid redox reactions emit energy above the frequency range capability of the human eye, and are therefore invisible.

Because flame is a product of a chemical reaction, the color of the flame is dependent on the heat produced as well as the efficiency of the combustion. The fewer solids produced, the bluer will be the flame. Solids tend to act as energy absorbers cooling both the reaction and reducing the amount of energy given off. The cooler the reaction, the lower the emitted frequency, the more yellow the flame. When the reaction temperature is very high the flame color approaches the blue end of the visible spectrum because the emissions are of a higher frequency.

A flame produced by burning propane with insufficient oxygen will be bright yellow and very smoky. The black smoke which is given off is unoxidized carbon—the solids in the flame. If more oxygen is provided in the form of air or even as pure oxygen, the temperature of the reaction will be increased, the flame temperature will be higher, and the flame color will become more blue. The smoke will be reduced.

There is at least one type of extremely high temperature fire that is invisible. This is the result of continuous oxidation of a limited flow of hydrogen in air such as might occur if a cryogenic hydrogen transporter accidentally sprang a small leak, and the escaping hydrogen were ignited by a spark or some other hot surface. The hydrogen in combination with the oxygen in the air would produce no solids, and because the reaction temperature is high the flame would be invisible, approaching the ultraviolet area of the spectrum. The reaction is written:

$$2H_2 + O_2 \longrightarrow H_2O + heat \qquad (136.6 \text{ kcal})$$

The fact that the flame is invisible in no way makes it less hazardous—quite the opposite. This fire would be extremely hazardous for two reasons: other flammable material or, worse, protective clothing could be ignited by inadvertent contact with the flame; and the reaction of hydrogen and oxygen is normally very explosive. Even a half pint of hydrogen, if mixed with the proper amount of air, produces a good bang. If more than a small leak occurs in a cryogenic transporter tank because of either secondary actions or melting of metal around the leak hole, a devastating explosion can be the result.

NONOXYGEN FLAME

There is another type of oxidation that can produce tremendous amounts of heat and visible flame with no oxygen present. The reactions are true redox reactions but the oxidizing and reducing agents are locked up in each molecule, needing only a small amount of trigger energy to initiate the reaction. Not unlike the sibling rivalry that exists in so many families, it doesn't take much to turn a peaceful, stable scene into a raging battle evolving considerable heat. The flash of a spark, a spurt of flame, or even the presence of certain metals called catalysts[1] can turn the coexisting oxidizing and reducing agents loose.

A popular rocket propellant which exhibits this characteristic is the non-hydrocarbon called hydrazine. Its formula is written N_2H_4 or, better, H_2NNH_2. Its structure is:

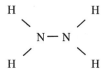

Hydrazine is normally a stable liquid at room temperatures but above 90 °F to 100 °F (32 °C to 38 °C), it becomes sensitive, bordering on the unstable. Triggered when either warm or cold, it will decompose very rapidly with the evolution of considerable energy. If hydrazine is sprayed onto platinum-bearing pellets or platinum screen, combustion begins at once. The platinum is a good catalyst for hydrazine. Cold hydrazine dripping onto wood will very soon oxidize the cellulose and initiate combustion.

STABILITY

Hydrazine is an example of a material which borders on the unstable. When the entire spectrum of chemical compounds is considered, we find a complete range of stabilities from total lack of activity with any other element or compound, to the extremely short-lived radioactive elements that decay and disappear in seconds.

A stable element or compound can lose its stability for what seems to be a variety of reasons. Hydrazine is relatively stable under impact or shock but it is sensitive to an increase in temperature. Nitroglycerine is relatively insensitive to temperature change but is very sensitive to shock. If, however, we consider only sensitivity to energy input, the apparent variety of reasons that appear to be reactions triggers, disappears. All of the reasons are forms of energy. The compounds react to the intensity as well as to the duration of that energy input. Temperature, impact, and spark are all forms of energy with varying degrees of intensity and duration. It is theorized that the characteristic stability or instabil-

---

[1]A catalyst is a substance that influences a chemical reaction, as to either speed or initiation, without itself undergoing a permanent change.

ity of an atom or molecule is inherent in the structure of the system. In atoms, stability may be due to a complete complement of outer shell electrons. In molecules instability is probably caused by the stresses on the linkages or bonds which exist between atoms. If the linkages are under great stress, it doesn't take much energy to snap them; energy is evolved, more bonds are broken, and the reaction becomes self-sustaining.

There are very few substances, other than the inert gases, that will not react with oxygen or that cannot be oxidized if placed under the proper conditions. Because of the stability of some elements or compounds, those proper conditions may be very high temperature or large amounts of electrical energy. For example, diamonds (carbon) can be burned with a resultant flame if the temperature of the diamond is raised above 1292 °F (700 °C).

At the opposite end of the stability scale are the substances that react seemingly with no provocation or energy trigger at all. Sodium, potassium, and calcium are so reactive that mere contact with air or water starts the reaction. The reaction is rapid and sufficiently exothermic to produce flame. These metals are such good reducing agents that when the supply of atmospheric oxygen is gone they will continue to produce flame by reducing the nitrogen from the air. Their high reactivity is the reason we do not use these metals as we do others, such as copper or iron.

SPONTANEOUS COMBUSTION

Spontaneous combustion is a term considered by many fire authorities to be a misnomer. They feel that the word spontaneous carries the connotation of a very short time period and therefore should be applied to very rapid combustion such as an explosion or detonation rather than to a very slow heat developing process.

However, in its rigorously correct usage the word carries no time concept. It does connote internal or self-contained drive to an end. There has become attached to the word a meaning which indicates something done as a momentary or impulsive act, without thought. Here again, while there is a shade of time involved in this meaning, the true indication is that of *internal forces acting without external influences.*

Spontaneous combustion is then a correct and excellent term to describe what happens in redox reactions that generate enough energy within themselves, and with no external influence, to raise the temperature of the mass to the ignition point. The reaction will take place only if certain factors or components are present.

While the fire component symbol has fuel, oxidizer, and energy in it, the spontaneous combustion symbol would look like this:

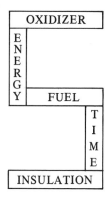

The major difference between simple fire initiation and spontaneous combustion is the addition of the two components *time* and *insulation*. The energy in this case is internal but it may be brought in from external sources such as microorganisms, as is the case with moist hay or cut grass fires. The most important element in spontaneous combustion is insulation; without it there can be no reaction. The insulation need not be of high quality; it need only fulfill the requirement that it retain more of the generated heat than it releases. Any closed space can provide the heat retention and so allow the temperature to rise above the ignition point.

Moist hay sitting in a stack in an open field rarely catches fire because of internally generated heat. That heat is certainly developed, but without an enclosure it escapes into the surrounding air. In a silo that same moist hay is almost a guaranteed barn burner. Anyone who doubts that moist hay or grass can generate heat should run a simple experiment. The next time the lawn is mowed, the clippings should be gathered up and put into a 20 to 30 gallon metal garbage can. Leave it for a few days in an open area, well away from any combustibles. Even after only a few hours the temperature change will be obvious to anyone touching the bottom of the container. In a few days the container will be hot on the bottom and lower sides. If the enclosure is of combustible material, as is a silo, a serious fire can result.

In this moist hay experiment, microorganisms feeding on the moisture and hay start and restart the redox reaction. That exothermic reaction then produces the heat which, if it does not escape, continues not only to accumulate but also to increase the reaction rate, building up more heat until the ignition point is reached.

Gasoline or solvent-soaked rags will react in a similar manner. This reaction, however, has no microorganisms involved. Most organic solvents and gasoline have notoriously low boiling points, and therefore vaporize very easily. The vapors, being a gas phase, have an extremely high surface area exposure where atmospheric oxygen molecules can readily react. The exothermic oxidation reaction, if enclosed, promotes heat buildup, developing a faster and faster reaction until the ignition temperature of the rag is reached.

OTHER METHODS OF IGNITION

While spontaneous ignition is a method of self-ignition requiring insulation, there are several external mechanisms which are able to provide the energy needed to start a redox reaction with or without insulation: an open flame (redox reaction); an electric spark (man made, lightning, or static electricity); mechanical friction; a hot surface; compression; impact (metal on metal, flint on steel, or crystal fracture); and optical (lensing).

OPEN FLAME

Open flame is a rather obvious method of raising the mass of a combustible substance to its ignition point. In short, one redox reaction can initiate another.

ELECTRIC SPARK

An electric spark is actually only a stream of electrons bursting from an area of high electrical pressure into an area of lower electrical pressure, as nature tries to restore its balance. The high electrical pressure exists because many extra electrons have been forced into the area by some chemical or mechanical action. The low electrical pressure area exists due to a lack of electrons. The higher the voltage difference between the two areas, the higher the electrical pressure; the higher the electrical pressure, the greater the gap-length the spark will jump. The visible manifestation of the electrons bursting from the higher pressure point to the lower occurs when the gap, whatever the material, breaks down or allows the electrons to flow from high point to low point. We call the visible manifestation a *spark*. Actually it is light given off by the ionized molecules of the material which have been excited enough by the electrical pressure to allow a flow of electrons. As the electrons flow from atom to atom at a speed of about 186,000 miles per second, the air around them is heated very rapidly. As this suddenly heated air expands it pushes against the cooler air surrounding it and a pressure wave is formed. This pressure wave is transmitted to other air molecules until it reaches our ears and we hear it as a zap for small spark gaps with minimum breakdown voltage, a crack for larger gaps and higher voltages, and just plain thunder if the pressure wave is generated by a tremendous spark breaking down a huge gap, that is, lightning.

This burst of energy is usually enough to initiate redox reactions, depending on the stability of substances in or near the gap and the amount of energy bridging the gap. In the case of lightning, the redox reaction results in atmospheric nitrogen being oxidized to nitrogen dioxide and nitrogen tetroxide.

Another form of electrical energy that is able to initiate redox reactions is static electricity. There is really nothing static, in the sense of stationary, about it. A better name for it is frictional electricity inasmuch as it is generated by rubbing two materials together. Of these two materials, one must be a good insulator and one a good electron donor. In the generation of this type of electricity, vigorous rubbing of one surface on the other causes the extraction of electrons from the surface of the donor material and the forcing of these electrons onto the surface of the insulator. An electrical pressure is therefore built up. When the electrical pressure is great enough it either discharges into the atmosphere or breaks down any convenient gap, arcing to a lower pressure conductor. Very high pressure differences have been generated by friction. This was all too often the case in older factories where a single electrical motor drove several machines by means of belts and pulleys moving at relatively high velocities. The friction of the moving belts has been known to generate good, hot sparks. When they occur near degreasing solvents, paints, or thinners, serious fires often result.

Frictional electricity is common with modern synthetic clothing. Even a short walk on synthetic carpeting can lead to tens of thousands of volts being built up on a person wearing synthetic clothing. Reaching for a metal solvent container under these conditions can lead to an explosion and fire.

MECHANICAL FRICTION

While rubbing certain materials together will produce frictional electricity, rubbing just about any solid substances together rapidly and with enough pressure will produce heat by means of mechanical friction. This heat, an indication of the energy imparted to the surface molecules by the abrading action, can grow under the proper conditions until the temperature of the material is above its ignition point.

It takes a person with know-how and experience to start a fire by rubbing two sticks together, but many a truck fire has been started in a locked brake drum when frictional heat built up to a point high enough to oxidize the oils and greases around the axle.

HOT SURFACE

Any surface containing sufficient energy in the form of heat can begin a redox reaction as long as the surface temperature is above the ignition point of the material that comes in contact with it. Any mass is an energy resevoir with its energy capacity directly proportional to the total mass as well as to the molecular structure. Therefore, if the mass is small or the energy is not being replenished or increased, the contacting material may act as a heat sink or robber and reduce the temperature of the mass below the ignition point of the contacting material. Thermodynamicists picture heat as always running down hill—always going from the warmer to the cooler object.

A complete range of conditions may result from hot surface contact depending on the materials of both surfaces and the amount of energy available. These conditions may vary from no visible change in the contacting material, through a range of scorching, to charring and combustion. The final result will, of course, depend on the total amount of heat available as well as the ignition point of the contacting surface. Dropping iron nails onto a hot surface will produce no visible effect. Dropping hot steel wool into pure oxygen will result in the complete combustion of the steel. Dropping nitroglycerin onto a surface at 300 °F (149 °C) results in nothing more than a boiling away of the liquid while dropping it onto a surface at 600 °F (316 °C) produces dramatic explosions. In between these two extremes is the case of dropping sugar onto the hot surface of a stove. Smoke and plenty of carbon is the usual end result.

When hot surfaces contain enough energy, they produce radiated energy in the form of light. It is possible then to approximate the temperature of glowing objects by their emitted color.

Color	Approximate Temperature °F*	Approximate Temperature °C*
dull red	930–1100	500–600
dark red	1110–1470	600–800
bright red	1470–1830	800–1000
yellow red	1830–2190	1000–1200
bright yellow	2190–2550	1200–1400
white	2550–2910	1400–1600

*In conversion from one scale to the other, all temperatures have been rounded.

COMPRESSION

The gas phase of many solvents can be ignited by a rapid compression of the enclosed volume. The diesel engine is built on this principle. The compression must be so rapid that the heat generated from the sudden forcing together of the molecules is not lost but actually raises the temperature of the compressed gas up to or above its ignition point. In order to have the compression stroke energy transferred into the gas phase in the form of a temperature rise, little or no leakage of that gas phase can be allowed around the compression medium. The compression medium may be a mechanical piston or an extremely high pressure shock wave trapping the gas phase in an atmospheric enclosure. The compression is said to be adiabatic if no heat is lost during the stroke. Practically, this is almost impossible.

While compression ignition is usually thought to be confined to diesel-type engines, it can and has been reported as having happened in buildings and even open air. In a devastating explosion in a refinery, the shock wave from the explosion slammed into a partially-filled tank of a highly volatile solvent. The shock wave front was traveling so fast it acted like an invisible piston crushing the tank walls rapidly, compressing the gas phase in the unfilled portion of the tank and raising it above its ignition point. This same sort of action has been reported as having happened outside of a container because the air molecules surrounding the gas cloud could not move out of the way fast enough and therefore acted like an enclosure.

IMPACT

Impact is merely a way of putting a large quantity of energy into a mass in a very short time. A swinging hammer or a falling weight can develop a great amount of kinetic energy if the hammer is swung fast enough or the weight falls far enough. Again, as in the case of compression ignition, the input energy must bring at least a part of the material up to its ignition point. Due to the nature of impact, this part may be just a single point. The millions of molecules in that single point, however, can react to the input energy and be pushed to or beyond their ignition temperature. Materials in the liquid state do not usually react to impact but there are notable exceptions; nitroglycerine is one of these. A great many solids are impact-sensitive. These solids are all crystalline materials. It is theorized that when a blow of sufficient kinetic energy fractures a few crystals along with their highly stressed bonds, oxidation is started, and a devastating reaction results.

OPTICAL

Each of the initiating systems discussed so far has been either a chemical or mechanical method of concentrating energy in the substance to be oxidized. No one will argue that light is not energy or that sunlight is not an abundant energy source. It has been estimated that millions and millions of calories of energy fall on the United States every day in the form of sunlight. If we concentrate just a small part of that energy by the optical means of a lens, under the proper conditions we can raise the temperature of a given point of a mass above its ignition point and so begin combustion. The concentrating lens need not be a ground, optical lens. In the proper shape and under the proper conditions any translucent material can concentrate the radiant energy from the sun and so initiate combustion. From experience, it would appear that the proper shape is just about any curved surface, and that the proper conditions are those that favor temperature increase. While insulation in the strict sense is not required, there should be no cooling effect from air currents. The substance receiving the concentrated energy must, of course, be combustible.

There has been a case reported in which a glass of water on a window shelf, concentrating the energy from the sun into an open container of paint rags, has been blamed for initiating a fire.

There are also those materials which, when placed in contact with each other, need only the slightest amount of optical energy to cause them to react. Hydrogen and chlorine can be mixed together in a dark room, and are quite stable in the darkness, but the slightest bit of sunlight, even without a lens, will trigger a reaction.

ABNORMAL COMBUSTION RATES

While combustion reactions can proceed at very slow rates (decay) or very fast rates (detonation), there are an infinite number of rates of reaction between these limits. In dealing with combustion, several terms are often used incorrectly. The correct terms, in the order of increasing rate of reaction, are:

Combustion: A rapid oxidation with the evolution of heat and light.

Deflagration: A very rapid oxidation with the evolution of heat and light, as well as the generation of a very low pressure wave.

Explosion: A very rapid oxidation with the evolution of considerable heat, accompanied by violent disruptive effects due to a medium-velocity shock wave. All reaction products are gaseous.

Detonation: An extremely rapid, almost instantaneous, oxidation reaction with the evolution of considerable heat, accompanied by a very violent disruptive effect and an intense, high-speed shock wave. All reaction products are gaseous.

DETONATION AND EXPLOSION

From the above definitions it can be seen that a detonation is an extremely rapid explosion which is in itself a very rapid combustion reaction. A detonation is then an ultrafast combustion. In his classic publication of 1883, "Sur la force des matieres explosives," the Parisian Berthelot considers every explosive reaction to be caused by an initial rise in temperature somewhere in or on an explosive substance. This rise can be caused by heat, shock, or friction, with the initial reaction causing an internal increase in energy, which is passed from particle to particle. Berthelot separates the propagation of the reaction into two classes, combustion and detonation. The more subtle, intermediate differences of deflagration and explosion were added later, although Berthelot recognized their existence.

His elaborate experiments led him to excellent conclusions, which have been proven to be correct in just about every detail. Berthelot concluded that if, by some energy input system, one small part of an explosive material is raised to its temperature of explosion, a shock wave is produced which raises the temperature of the adjoining parts. They then are raised to or above their explosion temperature, and they too react. This action is transmitted with a continually increasing velocity, a miniature chain reaction, throughout the whole mass.

The initiating energy may be any one of the several types discussed earlier. The total time from initiation to completed reaction depends on a characteristic of the material called **detonation velocity**. Both the container and the physical state, that is the compaction density, of the material affect the detonation velocity. In general, the stronger or more confining the container and the higher the density, the faster will be the propagation of the detonation through the material.

Not long after Berthelot's work an Englishman, Sir William Abel, did some excellent work to increase the knowledge of detonation propagation. Abel found that just about any rate of reaction could be produced by burning explosives in reduced pressure containers. He found that his explosives reacted quite differently under reduced pressure than they did at atmospheric pressure. He reasoned, logically, that the lower the surrounding pressure, the more easily the gaseous products of explosive decomposition could escape carrying with them the exothermic energy. Obviously, the resistance of the air molecules at normal pressure to the escape of the reaction product molecules depends upon the rate at which they are generated. The shock transmitted to the adjoining molecules during the propagation of the shock wave within the mass will be proportional to the pressure of the produced gases and the resistance to movement of the air molecules. Under extremely fast-moving shock wave conditions, the air molecules cannot get out of way fast enough and so themselves act like another container. Under reduced pressure they move more easily and therefore act less like a container. This lack of containment reduces the pressure, or at least does not let the pressure build up, with a resultant lack of increase in the reaction rate.

It would appear, therefore, that in order to produce a detonation, the initial rate of reaction must exceed a characteristic threshold. If it does not the explosive may burn slowly, burn rapidly, or just explode.

Temperature as well as confinement affects the threshold of a detonation reaction. The higher the temperature of the mass just before it is subjected to initiating energy, the more easily it will react. Nitroglycerine is a good example of how temperature affects the final rate:

Approximate Temperature		Action
°F	°C	
360	182	boils with yellow fumes
380	193	evaporates slowly
390	199	evaporates rapidly
420	216	deflagrates
440	227	deflagrates rapidly
465	241	detonates sometimes (residue)
495	257	detonates every time (residue)
510	266	detonates completely
550	288	detonates completely with flame

The statement has been made that nitroglycerine is so stable at room temperature that it is possible to snuff out a burning match by plunging it into the liquid. The originator of that statement, if he is still alive, should recall that nitroglycerine is far more sensitive to mechanical shock than to temperature. The very act of plunging the match into the liquid could set off a detonation.

Many materials that detonate do so even in the absence of an oxidizing agent. In effect, they carry their own oxidizing agents in the form of oxidizing atoms. These atoms are linked to the total molecule by highly stressed bonds. The more highly stressed the bonds, the more easily they are broken. In other words, the more highly stressed the linkages and the better the oxidizing atoms at gaining electrons, the more unstable is the substance. More often than not these substances are rich in nitrogen and oxygen. Nitrogen, with its multiple valence tendency, and oxygen, with its excellent electron-gaining potential, make fine components for explosive materials.

Modern materials made for detonation purposes, both large and small, tend to be complex organic, nitrogen-oxygen containing compounds. While these materials are tailor-made for optimum safety in handling, and yet are designed to explode or detonate with the proper initiation, they are based upon materials which were discovered because they had accidently exploded or detonated during someone's experiments. They produced their effect first; the why was discovered later.

A few of the more common organic nitrogen-oxygen explosives are:

Common Name	Formula
trinitrotoluene	$C_6H_2(NO_2)_3CH_3$
picric acid	$C_6H_2(NO_2)_3OH$
nitroglycerine	$C_3H_5(NO_3)_3$
tetryl	$C_6H_2(NO_2)_4NCH_3$
trinitroanisol	$C_6H_2(NO_2)OCH_3$

There are many inorganic compounds as well that will explode or detonate. These too, for the most part, contain nitrogen and oxygen atoms. Compounds of chlorine and oxygen called **chlorates**, when mixed with sulfur, phosphorus, and many organic compounds, are very explosive. They require very little frictional energy to initiate their high-speed reactions. Chlorates are, of course, very good oxidizing agents.

Explosive and detonation reactions have the common characteristic of producing large volumes of gaseous products and no solids whatever. The ammonium nitrate detonation reaction illustrates this:

$$2NH_4NO_3 \longrightarrow 4H_2O + O_2 + 2N_2$$

as does the nitroglycerine detonation reaction:

$$4C_3H_5(ONO_2)_3 \longrightarrow 12CO_2 + 10H_2O + 6N_2 + O_2$$

The shock wave produced by the sudden generation of those large volumes of hot gases has been measured at several hundred thousand pounds per square inch. Expansion ratios of 15,000 to 20,000 to one are not uncommon. Because of the high pressure front and the high temperatures, these gases tend to be very reactive. This characteristic leads to secondary reactions. In a sudden disaster the hot, highly reactive oxygen is available to begin the oxidation of normally stable compounds. The result is a new fire or at least the promotion of the one already in progress. Thus something as harmless as inorganic fertilizer in the form of ammonium nitrate can practically wipe out a city when the stability of that inorganic compound is upset and considerable oxygen evolved.

REVIEW QUESTIONS

1. Define combustion. How does it differ from oxidation?
2. Why is oxidizer the correct term rather than oxygen, when discussing fire in general terms?
3. What is an ignition point?
4. What is flame?
5. Explain why some flames are yellow and smoky while others are blue and clean.
6. What is a catalyst? Name three and their uses.
7. Why is stability so important when considering hazardous materials?
8. Define spontaneous combustion.
9. List the conditions needed for spontaneous combustion. Which is most important? Why?
10. Explain the mechanism of spontaneous combustion.
11. List five other methods of fire initiation. Explain how each operates.
12. What is frictional electricity? How can it initiate combustion?
13. What is compression ignition? Explain.
14. Distinguish between combustion, deflagration, explosion, and detonation.
15. Detail the mechanism of a detonation reaction.
16. How does atmospheric pressure promote explosive and detonation reactions?
17. Explain how some materials in the absence of oxygen or an oxidizing agent can explode or detonate.
18. What is a chlorate? Explain.
19. What is a common characteristic of explosive and detonation reactions? How does this promote their destructive force?
20. How does an explosion or detonation cause secondary combustion reactions?

Hazardous Liquids/ 6

WHAT IS A LIQUID?

Fire as energy is often just the ingredient needed to push a liquid over its threshold of stability and into the region of instability. The result may be toxic vapors, flame, deflagration, explosion, or even detonation.

A liquid is defined as that state of matter wherein the molecules are free to change their relative positions but yet not so free that they can completely separate from each other. They are held in any given area only by the container and their own cohesive forces. When enough energy is added to the liquid it becomes a gas and the cohesive forces are no longer strong enough to restrict free movement.

CHARACTERISTICS

BOILING POINT

Because liquids do not burn as liquids but must first vaporize, the way a liquid becomes a gas is of prime importance to a firefighter.

The transition from a liquid state to a gaseous state takes place when, under normal conditions of temperature and pressure, the higher energy molecules break free of the surface and so become unrestricted. This slow transition from liquid to gas is more commonly known as evaporation. The transition may also occur at a greater rate when energy is added by means of heat in some form. This rapid transition is more commonly called boiling.

The mechanism of boiling or evaporation is relatively simple in light of molecular theory. The average velocity at which the molecules are moving is dependent on the average energy. Some have more energy and move faster, some less energy and move slower. Those that come suddenly to the surface of the liquid have considerably less intermolecular force operating on them, and these higher velocity molecules with their greater energy will burst through the surface like a fullback breaking through a defensive line. Once free of the liquid, they become gas molecules due to the massive reduction in intermolecular forces. If the container is open, the molecules will eventually wander off into the atmosphere. If the container is closed, the molecules, now that they are in the gas phase, will bounce around off the vessel walls and off the other molecules that are also in the gas phase.

This action continues until the pressure, caused by yet other molecules that have joined the original ones, forces the gas molecules which collide by chance with the surface liquid molecules to be retrapped. Once again, they become part of the higher density material called a liquid. This condition or state, in which some of the molecules are breaking out of the liquid phase to become part of the gas phase, while others are at the same time reentering the liquid from the gas phase, is called a **state of equilibrium**. Note that this equilibrium is *not* a static state, but a highly dynamic one.

Where there is no enclosure, no gas pressure will build up above the liquid and, therefore, no equilibrium can be established. All of the escaping molecules will eventually become a gas and are free to escape into the atmosphere. The rate at which evaporation occurs varies with the temperature and the individual nature of the liquid.

Because it is always the higher velocity molecules which, coming to the surface, break free, the average velocity of the molecules making up the remainder of the liquid is lowered by the loss of the energy (heat) of these higher velocity molecules. This loss of heat then results in a lowering of the temperature of the liquid. By this mechanism evaporation results in cooling, whether it is water on the roof of a building or alcohol on the skin.

The property which describes the tendency of a liquid to evaporate is called **vapor tension**. Because it is difficult to measure vapor tension, it is determined indirectly by measurement of the vapor pressure of the liquid. The vapor pressure of a liquid is defined as the pressure of the gas phase of the liquid on the surface of the liquid at equilibrium.

Inasmuch as vapor pressure is related to molecular activity, the vapor pressure of a liquid can be increased by an increase in the temperature of the liquid:

more energy input = more molecular activity = higher vapor pressure

Because evaporation takes place completely from the surface of a liquid, and boiling is defined as a rapid vaporization beneath the surface of the liquid, they are closely related. Boiling could be considered as forced evaporation wherein, for a given pressure, the average energy of the liquid molecules is increased greatly so that they form gas phase pockets beneath the surface. Because the gas is less dense and the pressure within those pockets greater than the pressure on the surface of the liquid, these pockets or bubbles rise to the surface, breaking free. Looking at boiling from the pressure aspect, if the pressure on the surface of the liquid, and therefore on the whole container, is high enough, the gas pockets cannot form beneath the surface. There will be no boiling. To put it another way, as the pressure on its surface is raised, as it is with water in a household pressure cooker, the boiling point is also increased. Figure 6-1 shows this influence from 1 atmosphere through 10 atmospheres for water.

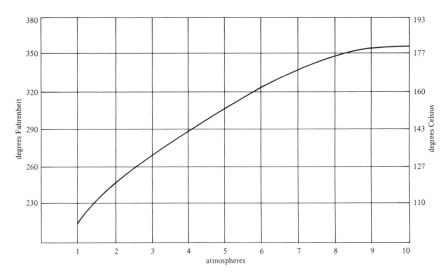

Figure 6.1. Effect of Pressure on Boiling Point (Water).

The reverse also holds true: when the pressure is lowered, as it is naturally at high altitudes, the boiling point decreases. For example, water boils at about 201 °F (94 °C) in Denver, Colorado but at 212 °F (100 °C) in New York City, or any sea level area.

Unless specifically stated the boiling point of a liquid is given at the standard pressure of 1 atmosphere or 14.7 psi (101 kilopascals). This is called the **normal boiling point**.

The boiling point of a liquid can be influenced in another way. By adding amounts of one liquid (or dissolved solid) to another, the boiling point of the solvent can theoretically be raised enough to bring a hazardous situation under control. The number of degrees of boiling point elevation depends on certain characteristics of both the solvent and the solute. Methyl alcohol mixes very well with water so it is theoretically possible to extinguish a fire of this type merely by hosing enough water into the alcohol to dilute it below its vaporization point. In actual practice, however, there are other important considerations that make the technique impractical. In order to dilute a tank of burning alcohol to reduce sufficiently its vaporization point, a large volume of water will be needed. This could result in tank overflow and spread of the flames before extinguishment. In addition, the heat of the fire can cause the water to boil, resulting in frothing and subsequent overflow.

While we have been discussing the boiling point of a single liquid, the question might well be asked, "What is the boiling point of a liquid that is made up of a physical blending of many different compounds, especially when they mix so well it is impossible to tell by simple observation that it is not a single liquid?"

The answer is that each liquid compound in that blend will have its own unique boiling point range. When the blend is heated and its temperature raised, the temperature of the liquid will continue to rise until the boiling point of the lowest boiling component or **fraction** is reached. At this point the temperature of the solution will hold relatively steady until all of that component has boiled away. It will then rise to the next highest boiling point and once again hold until that component has boiled away, and so on.

The process of using boiling points to separate mixtures of liquids, solids, and gases is called **distillation**. It is used in purifying water, refining crude petroleum, and other industrial applications.

Research on the relationship of boiling points and vapor pressures to carbon chain length indicates that the shorter chain length molecules have lower boiling points and therefore greater volatility. It would seem that the shorter the chain length, the more physically unstable is the liquid phase. This instability is the result of lighter, lower mass, molecules being able to react more easily to the lower energy input, increasing their velocity and their tendency to break away from the surface before the heavier ones.

FLASH POINT

Closely associated with the vapor pressure and boiling point is the flash point of a liquid. For fire safety personnel this is a very important characteristic.

In order to burn, liquids must first vaporize, and the vapors, uniting with atmospheric oxygen in unique ranges of combustible mixtures, can burn or even explode. The NFPA (National Fire Protection Association) defines a flash point as the lowest temperature at which a liquid gives off sufficient vapor to form an ignitable mixture with the air near the surface of the liquid or within the vessel containing the liquid. There is a characteristic minimum liquid temperature and range of vapor-to-air mixture at which some form of input energy, such as a flame or spark, can cause the vapor-air mixture to undergo an exothermic oxidation-reduction reaction and to produce a flash of energy. That minimum temperature is the flash point of the liquid.

There are several acceptable methods of determining this point but the most widely used technique seems to be the TAG, or Tagliabue, closed tester. There is also an open cup technique. When the flash point of a liquid is given it should be noted whether an open cup (oc) or closed cup (cc) was used. For more details on the systems, NFPA Bulletin No. 321 or ASTM (American Society for Testing and Materials) Bulletin D-56 should be consulted.

Obviously the direct relationship of vapor formation to flammability must involve external pressure as well as the availability of an oxidizer. Flash point determinations therefore are always made at a normal pressure of one atmosphere and standard air oxygen content. It is quite possible to lower the flash point of a liquid radically by reducing the surrounding pressure or increasing the available oxygen.

Conversely, as with the boiling point, the flash point of a liquid can be elevated by dilution, especially with water. The flash point of alcohol can be changed as follows:

Percentage of Alcohol	Percentage of Water	Flash Point	
		°F	°C
100	0	55	13
90	10	65	18
80	20	68	20
70	30	70	21
60	40	72	22
50	50	75	24
40	60	80	27
30	70	85	29
20	80	97	36
10	90	120	49
0	100	–	–

While the dilution of a hazardous liquid can make it less hazardous to handle, the evaporation of the diluting liquid can produce dangerous situations as the hazardous liquid becomes more concentrated once again.

The evaporation of liquid mixtures follows the same selective, fractional process as does the boiling point. A mixture of two liquids can change radically over a period of time if one of the liquids evaporates more quickly than the other. If the loss is in the diluent that is producing the safer mixture, the mixture can become a very unsafe liquid while still carrying the markings of a diluted, safe mixture. Because evaporation is normally a slow process, *time* is an important consideration but even more important is *containment.* In an enclosed container the vapor pressure will soon reach equilibrium and evaporation will stop. The greater the amount of liquid, the less room for vapor development, and therefore, the sooner equilibrium will be reached. Without confinement, there is no equilibrium, and a hazardous condition can develop.

This is precisely the reason that flammable liquids should be kept in closed containers with tight closures. It is also the reason that full containers are less hazardous than empty ones: less vapor = less risk of reaching and igniting at the flash point.

FIRE POINT

Another characteristic of liquids of importance to fire safety personnel is the fire point. While a liquid that receives sufficient input energy at its flash point will produce a momentary flash, no continuous combustion results. The energy output is almost instantaneous. That is not to say that ignition at the flash point cannot produce a disaster. If the volume of vapor-air mixture is enough, the energy output can be devastating. However, with the liquid a few degrees above the flash point, enough vapor can be produced so that the mixture is capable of developing enough energy for the reaction to become self-sustaining when ignited from an outside source. This is the fire point.

AUTOIGNITION POINT

At temperatures well above the fire point it is possible for enough energy to collect in the vapor-air mixture to self-ignite. This point, when it exists, is called the autoignition point or autoignition temperature. Under laboratory conditions it is defined as the lowest temperature to which a closed container must be heated so that the liquid under test, when injected into that container, will undergo combustion without any additional, external ignition source. It is, therefore, the lowest temperature at which a liquid will not only vaporize, but at which sufficient energy remains to promote a combustion reaction. No outside ignition is needed.

Delayed ignitions are normal with this test and so no time limitation is prescribed as a qualification between injection and ignition. The standard method for determining this characteristic is given in ASTM Bulletin D-2155, entitled "Autoignition Temperature of Liquid Petroleum Products."

In summary, then:

Name	Relative Temperature	Output	Self Sustaining?	Ignition
flash point	lowest	flash	no	external
flame point	medium	sustained	no	external
autoignition point	highest	continuous	yes	internal

VAPOR DENSITY

The vapor state is an intermediate stage between the liquid and gas states wherein liquid droplets or molecular clusters exist usually in an equilibrium with individual molecules (gas). The surface area is much larger than in the liquid state but nowhere near as great as in the gas phase. Because of the greater surface area, reaction rates can be very fast.

Vapor density is also important to fire safety personnel because it is an indication of whether the vapors given off by an evaporating or boiling liquid will rise or fall to the ground.

In fire chemistry, vapor density is usually the comparison or ratio of a weight of a volume of air to an equal weight of the same volume of a vapor. It can be considered similar to the specific gravity of a liquid except in this case the standard is normal air composition rather than pure water.

As with specific gravity, the standard (air) is taken as 1. Dimethyl ether, for example, with a vapor density of 1.59 is therefore 1.59 times heavier than air. In practical terms this indicates that these ether vapors are heavier than air and will therefore drift downward rather than upward. Actual vapor densities will always be lower than the listed theoretical because in actual cases vapors will be a mixture of the liquid vapor and air. Depending on the actual mixture ratio, the practical vapor density may be anywhere between 1 (air) and the theoretical vapor density for that liquid.

The theoretical vapor density of any substance may be found by dividing the molecular weight (usually in grams) by the total molecular weight of air (in grams). Keeping in mind that air is a mixture, the number 29 can be used as its molecular weight.

Methyl alcohol has the formula CH_3OH. Therefore the molecular weight is:

$$
\begin{array}{lll}
C = 12, & \text{so} & 12 \times 1 = 12 \\
H = 1, & \text{so} & 1 \times 4 = \underline{4} \\
O = 16, & \text{so} & 16 \times 1 = \underline{16} \\
\end{array}
$$

$$\text{Total } 32$$

According to the formula,

$$\text{vapor density} = \frac{\text{molecular weight of liquid}}{29 \text{ (molecular weight of air)}}$$

Therefore, the vapor density of methyl alcohol is:

$$\text{V.D.} = \frac{32}{29} = 1.1$$

It should be noted that of over 300 common liquids studied in relation to vapor density, only one, hydrocyanic acid, had a vapor density less than one. In short, the vapor density of liquids is usually heavier than air.

SOME SPECIFIC LIQUIDS

ALDEHYDES

These organic compounds are characterized as strong irritants to the body tissues, especially those in the eyes, nose, and throat. The lower carbon chain aldehydes are quite soluble in water, the higher carbon chain aldehydes much less so.

Formaldehyde, CH_2O

Formaldehyde in its pure form is really a gas; however, many commercial solutions of the material exist carrying the name of the pure compound. As an absolutely pure substance it is rare because it tends to polymerize. Common solutions range from 35 to 50 percent formaldehyde by weight with various solvents such as methanol and water. In these solutions it is a colorless liquid with a heavy aroma, having the following characteristics:

> specific gravity: 1.0 (approximately)
> boiling point: 210 °F (99 °C)
> flash point: 190 °F (88 °C)
> autoignition point: 800 °F (427 °C)
> ignition limits: upper 75%
> lower 7% (by volume)

In a fire situation, gaseous formaldehyde will be produced as soon as the solution begins to heat up. Toxic concentrations can build up very rapidly, especially in closed areas. Fire potentials with this material are only medium, but explosions have been reported where solutions have been heated well above their flash points.

Fire may be extinguished with dry chemicals, foam, or carbon dioxide.

Acetaldehyde, CH_3CHO

Acetaldehyde is a clear, colorless liquid, also known as ethyl aldehyde and acetic aldehyde, and has a heavy, fruit-like odor and the following characteristics:

> specific gravity: 0.8
> boiling point: 69 °F (21 °C)
> flash point: −36 °F (−38 °C) (cc)
> autoignition point: 365 °F (185 °C)
> ignition limits: upper 59%
> lower 4% (by volume)

This compound is a strong local irritant as well as a narcotic. It is a very hazardous substance when exposed to flame or any high energy input, reacting very rapidly with most oxidizing agents. It has been known to explode under these circumstances, and therefore must be treated as a very dangerous material.

Fire extinguishment should be attempted with alcohol foam, carbon dioxide, or dry chemicals.

Benzaldehyde, C_6H_5CHO

Sometimes known as synthetic or artificial almond oil, oil of almonds, and benzoic aldehyde, benzaldehyde is a colorless liquid with the following characteristics:

> specific gravity: 1.1
> boiling point: 354 °F (179 °C)
> flash point: 149 °F (65 °C)
> autoignition point: 378 °F (192 °C)
> ignition limits: not available

While this material will react with oxidizing materials, it does not do so very easily. Its high boiling, flash, and autoignition points indicate that it is a low fire hazard.

Fires may be extinguished with dry chemicals, alcohol foam, or carbon dioxide.

Butyraldehyde, $CH_3CH_2CH_2CHO$

Also known as butyric aldehyde, normal butyl aldehyde, and butanal, butyraldehyde is a colorless liquid with the following characteristics:

> specific gravity: 0.8
> boiling point: 168 °F (76 °C)
> flash point: 20 °F (7 °C) (cc)
> autoignition point: 446 °F (230 °C)
> ignition limits: not available

The low flash point of this substance makes it a high fire hazard. It will react readily with oxidizing materials but it is *not* prone to spontaneous heat generation or ignition.

Fires have been extinguished with dry chemicals, alcohol foam, or carbon dioxide.

Paraldehyde, $\overline{OCH(CH_3\ OCH(CH_3)OCCH_3}$ (cyclical)

Also known as para-aldehyde, as well as several other similar names, paraldehyde is a colorless liquid with the following characteristics:

> specific gravity: 1.0
> boiling point: 256 °F (124 °C)
> flash point: 96 °F (36 °C) (oc)
> autoignition point: 460 °F (238 °C)
> ignition limits: upper not available
> lower 1.2% (by volume)

This is a very dangerous substance from two aspects: it is a drug that has caused death from respiration of its vapors; and it can explode when exposed to enough heat or flame. Furthermore, it produces the highly toxic gases mentioned previously when heated to decomposition. It will react rapidly with oxidizing materials.

Fires can be extinguished with alcohol foam, dry chemicals, or carbon dioxide.

Acrolein, CH$_2$CHCHO
(unsaturated between C1 and C2)

Acrolein is also known as propenal, acraaldehyde, allyl aldehyde, and acrylic aldehyde, and is a straw-colored liquid with a very unpleasant aroma and the following characteristics:

specific gravity:	0.8
boiling point:	127 °F (53 °C)
flash point:	−14 °F (−26 °C)
autoignition point:	unstable, about 530 °F (277 °C)
ignition limits:	upper 30% lower 3% (by volume)

Acrolein has its own built-in warning system. Its very unpleasant aroma and its attack on the tear ducts and membranes of the eyes make it difficult to ignore. A maximum of 0.1 part per million parts of air is considered the threshold limit for safe respiration. Acrolein is also a definite fire hazard and is subject to self-heating. It will, moreover, react with oxidizing materials, and, while decomposing, emits very toxic fumes.

Alcohol foam, dry chemicals, or carbon dioxide may be used to extinguish acrolein fires.

Crotonaldehyde, CH$_3$CHCHCHO
(unsaturated between C2 and C3)

Also known as crotonic aldehyde, 2-butenal, propylene aldehyde, and β-methylacrolein, crotonaldehyde is a clear, colorless liquid with a strong aroma and the following characteristics:

specific gravity:	0.9
boiling point:	219 °F (104 °C)
flash point:	55 °F (13 °C)
autoignition point:	450 ° F (232 °C)
ignition limits:	upper 16% lower 2% (by volume)

While crotonaldehyde is not subject to self-heating, it is nevertheless a very dangerous and flammable material. It attacks the eyes, usually resulting in copious tears. It has been known to burn the cornea, leaving permanent damage. It will react quickly with oxidizers, and therefore must be kept away from flames.

Dry chemicals, alcohol foam, or carbon dioxide will extinguish crotonaldehyde fires.

Furfural, OC_4H_3CHO
(cyclic and unsaturated between C1 and C2, C3 and C4)

Furfural, also known as fural(e), 2-furaldehyde, furfuraldehyde, furol, and furancarbonal, is a yellow to straw-colored liquid with the pleasant smell of almonds. It has the following characteristics:

specific gravity:	1.2
boiling point:	323 °F (162 °C)
flash point:	140 °F (60 °C)
autoignition point:	600 °F (316 °C)
ignition limits:	upper 19% lower 2% (by volume)

The cyclic molecular structure of this material makes it unique among the commonly known aldehydes. Fumes from furfural will attack the eyes, but its high boiling point reduces its vapor production. While it is not subject to self-heating, it is nevertheless a moderate fire hazard and does react with oxidizers. It has been reported that furfural reacts explosively with inorganic acids and bases.

Fires can be extinguished with water, dry chemicals, alcohol foam, or carbon dioxide.

Propionaldehyde, CH_3CH_2CHO

Propionaldehyde is also known as propyl aldehyde, propanal, and propylic aldehyde, and is a colorless liquid with a heavy, overpowering aroma and the following characteristics:

specific gravity:	0.8
boiling point:	120 °F (49 °C)
flash point:	17 °F (8 °C) (oc)
autoignition point:	405 °F (207 °C)
ignition limits:	upper 17% lower 3% (by volume)

The relatively low boiling and flash points of propionaldehyde make it an ideal candidate for fire problems. It reacts quite rapidly with oxidizers and so has been accused as the cause of many serious fires.

Fires have been extinguished with dry chemicals, alcohol foam, and carbon dioxide.

ALCOHOLS

The alcohols, as a group, have the highest boiling points of flammable liquids. This is not to say they are not hazardous; they burn with ease and can produce fires that are very difficult to extinguish.

The concentration of alcohol, usually in water, is given as the proof, which is about twice the percentage of alcohol in the water. For example, 180 proof alcohol is 90 percent alcohol and 10 percent water.

The four most popular alcohols in common usage are methyl, ethyl, isopropyl, and phenol.

Methyl Alcohol, CH_3OH

Also known as wood alcohol and methanol, methyl alcohol is a clear, thin liquid with the following characteristics:

specific gravity:	0.8
boiling point:	149 °F (65 °C)
flash point:	52 °F (11 °C)
autoignition point:	867 °F (464 °C)
ignition limits:	upper 36% (by volume)
	lower 7%

Methyl alcohol is a narcotic, as well as an irritant that attacks the mucous membranes and eyes. It is cumulative in the body because it is eliminated very slowly.

While the explosion hazard with this substance is only medium, it is a very high fire hazard. Although without spontaneous ignition problems, it will react very rapidly with oxidizing materials.

Fire with methyl alcohol should be fought with dry chemicals or carbon dioxide.

Ethyl Alcohol, C_2H_5OH

Ethyl alcohol, also known as spirits of wine, spirit of alcohol, methyl carbinol, and ethanol, is a clear, colorless liquid with a unique fragrance and the following characteristics:

specific gravity:	0.8
boiling point:	173 °F (78 °C)
flash point:	55 °F (13 °C)
autoignition point:	793 °F (423 °C)
ignition limits:	upper 19% (by volume)
	lower 4%

Vapor concentrations of 5000 to 10,000 parts per million can result in eye and mucous membrane irritation as well as respiratory problems. Ethyl alcohol differs from methyl in its chemistry in the body, in that the human body can quickly oxidize ethyl alcohol to produce carbon dioxide and water. There are no cumulative effects over an extended period of time, as with methyl alcohol.

Like methyl alcohol, however, it presents a great fire hazard. While it does react quickly with oxidizing materials, it does not present a spontaneous ignition problem. Even when exposed to open flame, the explosion hazard is only moderate.

Ethyl alcohol fires should be extinguished with alcohol foam, dry chemicals, or carbon dioxide.

Isopropyl Alcohol, C_3H_7OH

Isopropyl alcohol, also known as rubbing alcohol, isopropanol, dimethyl carbinol, and secondary propyl alcohol, is a clear, water white liquid with the following characteristics:

specific gravity:	0.8	
boiling point:	180 °F (82 °C)	
flash point:	53 °F (12 °C)	
autoignition point:	750 °F (399 °C)	
ignition limits:	upper 12%	
	lower 2%	(by volume)

As with all alcohols, isopropyl is narcotic, but only in high concentrations. It will cause eye, nose, and throat damage even with concentrations as low as 400 parts per million (ppm). The body seems to treat it like ethyl alcohol as far as elimination is concerned, with tolerances varying widely with different individuals. There are reports of serious medical problems arising with as few as 150 drops of isopropyl alcohol taken internally.

It will react rapidly with oxidizing materials but does not present a spontaneous ignition threat. It is a fire hazard, as are all alcohols.

Isopropyl alcohol fires should be extinguished with dry chemicals or carbon dioxide.

Phenol

Phenol is an aromatic alcohol, also known as carbolic acid, phenylic acid, and phenic acid. In its pure state it is a solid at room temperature, and therefore is covered under hazardous solids.

Denatured Alcohol

Also called denatured spirits, denatured alcohol is not chemically distinct but rather a mixture of a poison (denaturant) and an alcohol. More often than not the alcohol will be ethyl and the denaturant methyl. Sulfuric acid is also used as a denaturant. The chemical characteristics will vary with the substance used as the denaturant.

As far as fire and explosive hazards are concerned, denatured alcohol should be treated like ethyl alcohol.

Spirits of . . .

There are many solutions of an alcohol with a liquid or a solid used in industry and the home that carry the name "spirits of . . ." The hazards associated with each mixture will depend on the concentration of solute as well as its chemical nature. Spirits of nitroglycerin can be almost harmless until the alcohol evaporates and leaves pure nitroglycerin as a residue. It then becomes a primary potential hazard.

A note of caution: not all "spirits of . . ." are alcohol solutions. Spirits of turpentine and mineral spirits, to name but two, are not alcohol solutions, but are basically turpentine. There are others.

Some other liquid alcohols are:

Name	Specific Gravity	Boiling Point		Flash Point		Autoignition Point		Upper Ignition Limit (%)	Lower Ignition Limit (%)
		°F	°C	°F	°C	°F	°C		
amyl alcohol	0.8	280	138	91	33 (cc)	572	300	9	1
butyl alcohol	0.8	243	117	84	29	690	366	11	1
furfuryl alcohol	1.1	340	171	167	75 (oc)	915	491	15	2
cyclohexanol[1]	0.9	323	162	154	68 (cc)	572	300	–	–
diacetone alcohol	0.9	334	168	150	66	1115	602	7	2
octyl alcohol	0.8	360	182	178	81	–	–	–	–

[1] Exists as both a liquid and a solid.

Attempt to extinguish liquid alcohol fires with alcohol-type foam or dilution with water, but the latter only with care.

AMINES

The amines, as their name suggests, are related to ammonia. The lower molecular weight amines tend to be gases at normal temperatures and pressures, but there are many liquid amines that are quite volatile, producing an odor that can only be described as a combination of ammonia and rotten fish.

Because of high volatility and toxic vapors, the amines are all extremely hazardous. Ordinary gas masks are not generally effective and therefore only self-contained breathing apparatus should be used.

Aniline, $C_6H_5NH_2$

Also known as aniline oil, phenylamine, and aminobenzene, aniline is a slightly straw-colored to clear, oily liquid, with the following characteristics:

specific gravity:	1.0
boiling point:	364 °F (184 °C)
flash point:	158 °F (70 °C)
autoignition point:	1415 °F (768 °C)
ignition limits:	upper (not available)
	lower 1% (by volume)

The great danger with aniline is not so much from the fire hazard as from the toxic vapors it produces when heated to decomposition. As a liquid it can produce skin rashes, serious in some cases. Therefore, the liquid, as well as the fumes, should be vigorously avoided.

In fighting an aniline fire, self-contained masks must be used. Such fires may be extinguished with dry chemicals, alcohol foam, or carbon dioxide.

Monomethylamine–Aqueous, $CH_3NH_2 \cdot H_2O$

Normally a gas, monomethylamine is often dissolved in water, up to about 35 percent by volume. The word "aqueous" is then added to the name.

Carbon dioxide, alcohol foam, or dry chemicals can be used to extinguish fires involving this solution.

Ethylamine, $CH_3CH_2NH_2$

Also known as ethyl amine, aminoethane and monoethylamine, ethylamine is a colorless liquid often found in 70 percent solutions with water, and has the following characteristics:

specific gravity:	0.7
boiling point:	88 °F (31 °C)
flash point:	less than 0 °F (−18 °C)
autoignition point:	723 °F (384 °C)
ignition limits:	upper 14%
	lower 4% (by volume)

This is a typical amine with all of the related toxicology. In addition, it is a very dangerous material that reacts vigorously with heat or flame to produce fires that are very difficult to extinguish. It will also readily react with oxidizing materials.

Dry chemicals, alcohol foam, or carbon dioxide can be used to fight ethylamine fires.

Ethylenediamine, $H_2NCH_2CH_2NH_2$

Also known as 1,2 diaminoethane, ethylenediamine is a very volatile, colorless liquid with the following characteristics:

specific gravity:	0.9

boiling point:	242 °F (117 °C)
flash point:	93 °F (34 °C) (cc)
autoignition point:	725 °F (385 °C)

Compared to other amines, this is a mild member of the family that tends to pick up moisture from the air and so dilute itself. It is only a medium fire hazard.

In spite of its relatively low fire danger, ethylenediamine is a full member of the amine family as far as skin irritation and the dangers of vapor inhalation are concerned.

Ethylenediamine fires can be extinguished with dry chemicals, carbon dioxide, or alcohol foam.

Triethylamine, $(CH_3CH_2)_3N$

The tertiary amine triethylamine is a very dangerous, colorless liquid that produces vapor that has caused inhalation problems with as few as 30 parts per million in air. It has the following general characteristics:

specific gravity:	0.7
boiling point:	193 °F (89 °C)
flash point:	20 °F (7 °C) (oc)
autoignition point:	(not available)
ignition limits:	upper 8%
	lower 1% (by volume)

This compound is dangerous not only because of its fire potential but also as a material that produces highly toxic vapors. Absorption through the skin is quite possible with triethylamine, and permanent sensitization has been reported.

Triethylamine reacts very quickly to input energy in any form, and will readily react with oxidizing materials.

Self-contained breathing apparatus and complete skin coverage is absolutely required around this compound.

Dry chemicals, alcohol foam, or carbon dioxide can be used to fight triethylamine fires.

ALCOHOL-AMINE HYBRIDS

In the amine group there is a series of hybrids that are compounds, having the alcohol as well as the amine linkage. The more common ones are ethanolamine, diethanolamine, and triethanolamine, with one, two, or three alcohol chains, respectively, attached to the nitrogen atom of the amine.

Ethanolamine, $HOCH_2CH_2NH_2$

Also known as monoethanol amine and 2-aminoethanol, ethanolamine is a color-less liquid with a distinct ammonia odor. It has the following characteristics:

specific gravity:	1.0
boiling point:	339 °F (171 °C)
flash point:	200 °F (93 °C) (oc)

The safe maximum allowable concentration of ethanolamine in air is only 3 parts per million for long term exposure. However, because there is much to be learned about the toxicity of this substance every precaution should be taken for even short term exposure—it *is* an amine.

Fires involving ethanolamine may be fought with dry chemicals, alcohol foam, or carbon dioxide.

Diethanolamine, $(HOCH_2CH_2)_2NH$

Diethanolamine is a secondary amine, also known as di (2-hydroxyethyl) amine. It is a straw-colored, thick liquid with the following characteristics:

specific gravity:	1.1
boiling point:	516 °F (269 °C)
flash point:	305 °F (152 °C) (oc)
autoignition point:	1224 °F (662 °C)

The fire hazard with this compound is very low, as is its irritability and inhalation toxicity. It has been known to react with oxidizing materials, but not with the ease of most amines.

Extinguish these fires with dry chemicals or carbon dioxide.

Triethanolamine, $(HOCH_2CH_2)_3N$

Triethanolamine, a tertiary amine, is a light, straw-colored liquid with the following characteristics:

specific gravity:	1.1
boiling point:	680 °F (360 °C)
flash point:	355 °F (179 °C) (cc)

Triethanolamine presents only a slight fire hazard, and a very low hazard for skin absorption. However, when heated sufficiently, the decomposition products are very toxic. It will react very rapidly with oxidizing agents.

Triethanolamine fires have been extinguished with dry chemicals, carbon dioxide, or water. Frothing must be avoided when water is used.

Other Liquid Amines

A few of the more common liquid amines other than those already detailed are:

Name	Specific Gravity	Boiling Point		Flash Point		Autoignition Point		Upper Ignition Limit (%)	Lower Ignition Limit (%)
		°F	°C	°F	°C	°F	°C		
propyl amine	0.7	120	49	−35	−37	604	318	10	2
allyl amine	0.8	128	53	−20	−29	705	374	22	2
n-butyl amine	0.8	171	77	45	7	594	312	10	2
n-amyl amine	0.8	219	104	45	7	−	−	−	−
methylaniline[1]	0.8	303	151	120	49	991	533	−	−
toluidine[1]	1.0	398	203	−	−	−	−	−	−
hexyl amine	0.8	270	132	85	29	−	−	−	−

[1] As with most aniline derivatives, this compound is quite hazardous as far as vapors and skin contact are concerned.

Amine fires are best extinguished by remote fire-fighting methods using carbon dioxide or alcohol foam.

ETHERS

Because ethers are characterized by low boiling points and high flammability, they are very hazardous in any fire situation. Furthermore, most ethers have a specific gravity less than 1.0, that of water, indicating that they will float fire on its surface. There are very few exceptions.

A note of caution concerning ether nomenclature: the ether linkage (—O—) carries a chain on both sides of that oxygen atom. For this reason when the chains are identical the prefix di- should appear in the name. In actual practice, however, the di- is often omitted. For example, dimethyl ether is sometimes written methyl ether with the understanding that an ether must have a carbon chain on both sides of the linkage, and therefore it must be methyl if not otherwise noted.

Aliphatic ethers are quite volatile and therefore extremely flammable. The lower carbon chain ethers have very low flash points and relatively heavy vapors. One of the more hazardous traits of the ethers is the tendency to react chemically with atmospheric oxygen to form very unstable peroxides. These peroxides are so unstable, in fact, that the slightest jar will often result in an explosion.

Ethers, especially the lower molecular weight compounds such as methyl and ethyl ether, must be kept away from all abnormal energy inputs. Not only open flames but even hot plates have been reported as being the ignition source for their vapors. Ethers must also be kept isolated from strong oxidizing materials. It is good practice to keep them isolated in cool, well-ventilated areas, where their containers are electrically grounded, to prevent static electricity accumulation or a spark of any kind.

Dimethyl Ether, CH_3OCH_3

Also called methyl ether and methyl oxide, dimethyl ether is a highly volatile and colorless liquid that is an extreme hazard when exposed to the slightest external energy input, including the heat from normal room temperatures. It must be kept pressurized to keep it from vaporizing. It has the following characteristics:

specific gravity:	0.67
boiling point:	$-11\,°F\,(-24\,°C)$
flash point:	$-42\,°F\,(-41\,°C)\,(cc)$
autoignition point:	$660\,°F\,(349\,°C)$
ignition limits:	upper 18%
	lower 3% (by volume)

Dimethyl ether vaporizes with extreme ease to produce very dangerous conditions in an area with even the slightest confinement for the vapors. Laboratory cases have been reported of a carelessly left open container producing vapors overnight, resulting in an explosion and fire the next morning when an electric switch was thrown to turn on the laboratory lights.

Because its boiling point is so low, dimethyl ether is usually shipped as a pressurized gas in heavy cylinders.

Attempt to extinguish fires with great care, using dry chemicals or carbon dioxide. If dimethyl ether is appearing as a gas, stop the gas flow first.

Diethyl Ether, $CH_3CH_2OCH_2CH_3$

Sometimes called ether, anesthetic ether, ethyl ether, and ethyl oxide, diethyl ether is a clear volatile liquid with the following characteristics:

specific gravity:	0.7
boiling point:	$94\,°F\,(34\,°C)$
flash point:	$-49\,°F\,(-45\,°C)$
autoignition point:	$356\,°F\,(180\,°C)$
ignition limits:	upper 48%
	lower 2% (by volume)

While diethyl ether is not especially reactive chemically, it will react with many oxidizing agents when influenced by small amounts of heat. It is therefore extremely hazardous when exposed to heat, flame, or a spark.

It can be very toxic when inhaled in a nonmedically controlled situation. Deaths have been reported from heavy and/or continual exposure.

Diethyl ether fires should be extinguished with alcohol foam, carbon dioxide, or halons.

Isopropyl Ether, $(CH_3)_2CHOCH(CH_3)_2$

Also called diisopropyl ether and 2-isopropoxy propane, isopropyl ether is a colorless liquid with the following characteristics:

specific gravity: 0.7
boiling point: 155 °F (68 °C)
flash point: −18 °F (−28 °C)
autoignition point: 830 °F (443 °C)
ignition limits: upper 21%
lower 1%   (by volume)

Like other ethers, isopropyl ether is a highly explosive substance when subjected to heat or an input energy impulse. Isopropyl ether will not spontaneously ignite at normal temperatures, but it will react violently with oxidizing materials.

Isopropyl ether generates highly toxic gases when it decomposes. It is reported that it will explosively react to impact. While it is only a moderate hazard on inhalation, it is a strong irritant and has been known to cause permanent injury after even short exposures. It seems to attack the skin and/or mucous membranes. It is, of course, a narcotic, but only in high concentrations.

Alcohol foam or carbon dioxide may be used to extinguish isopropyl ether fires.

Diethylene Glycol Diethyl Ether, $CH_3(CH_2OCH_2)_3CH_3$

Diethylene glycol diethyl ether, also known as ethylene glycol ethyl ether, is a colorless liquid that is soluble in organic solvents as well as water. It has the following characteristics:

specific gravity: 0.9
boiling point: 372 °F (189 °C)
flash point: 180 °F (83 °C) (oc)

In this chemical combination of glycol and ether, the glycol characteristics seem to predominate with resultant high boiling and flash points. It is, therefore, only a medium fire hazard, and then only when exposed to high input energy impulses.

Use alcohol foam or halons to attempt to extinguish fires.

Tetrahydrofuran, $OCH_2CH_2CH_2CH_2$ (cyclic)

Tetrahydrofuran is a cyclic ether also known as cyclotetramethylene oxide, diethylene oxide, and tetramethylene oxide. It is a colorless liquid with the following characteristics:

specific gravity: 0.9
boiling point: 150 °F (60 °C)
flash point: 1 °F (−17 °C)
autoignition point: 610 °F (321 °C)
ignition limits: upper 12%
lower 2%   (by volume)

Tetrahydrofuran is a very dangerous substance that will readily react with oxidizing materials to produce very persistent flames. In addition, when burning or even just heated to decomposition, it produces highly toxic gases. It also reacts with air to form unstable, explosive peroxides.

Tetrahydrofuran is a very strong skin and lung irritant that attacks the mucous membranes and eyes. It is a narcotic in high concentrations.

Attempt to fight fires with alcohol-type foam.

Other liquid ethers are:

Name	Specific Gravity	Boiling Point		Flash Point		Autoignition Point		Upper Ignition Limit (%)	Lower Ignition Limit (%)
		°F	°C	°F	°C	°F	°C		
anisole	1.0	309	154	125	52	–	–	–	–
divinyl ether	0.8	102	39	–22	–30	680	360	27	2
diallyl ether	0.8	202	94	20	7 (cc)	–	–	–	–
diamyl ether	0.8	338	170	135	57 (oc)	340	171	–	–
dibutyl ether	0.8	285	141	78	26	382	194	8	1
2,2 dichloroethyl ether[1]	1.2	353	178	131	55 (cc)	696	369	–	–
1,4 dioxane	1.0	214	101	54	12 (cc)	356	180	22	2

[1] Reacts with water to generate very toxic fumes.

Use dry chemicals, alcohol foam, or carbon dioxide to extinguish liquid ether fires. However, boiling may displace foam blanket.

KETONES

Ketones are chemically similar to·aldehydes, but the small structural difference makes a large difference in their properties and characteristics. While the aldehydes are generally toxic and irritants, no such generalization can be made for the ketones.

Acetone, CH_3COCH_3

Also known as propanone and dimethyl ketone, acetone is a colorless, water soluble liquid with the following characteristics:

specific gravity: 0.8

boiling point: 134 °F (57 °C)

flash point: 0 °F (–18 °C) (cc)

autoignition point: 1000 °F (538 °C)

ignition limits: upper 13% / lower 3% (by volume)

The low flash point of acetone makes it a fire hazard. In addition, it is a common solvent that has gained wide popularity, especially in the plastics industry. It will explode when the vapors are exposed to spark or flame and will react with oxidizing materials without very much input energy.

Extinguish fires with alcohol foam or carbon dioxide.

Methyl Ethyl Ketone, $CH_3COCH_2CH_3$

Also known as ethyl methyl ketone, MEK, 2-butanone and butanone, methyl ethyl ketone is a water soluble, colorless liquid, and an excellent and common solvent with the following characteristics:

specific gravity:	0.8
boiling point:	175 °F (79 °C)
flash point:	22 °F (−6 °C) (oc)
autoignition point:	960 °F (516 °C)
ignition limits:	upper 10%
	lower 2% (by volume)

Methyl ethyl ketone is a very dangerous material that can react with most oxidizing materials to produce disastrous fires. It is a moderate explosive with the very narrow limits of its ignition points needing only a small spark or flame to start the reaction.

Extinguish methyl ethyl ketone fires with dry chemicals, alcohol foam, or carbon dioxide.

Diethyl Ketone, $C_2H_5COC_2H_5$

Also known as propione, ethyl propionyl, metaacetone and 3-pentanone, diethyl ketone is a colorless liquid with the following characteristics:

specific gravity:	0.8
boiling point:	214 °F (101 °C)
flash point:	55 °F (13 °C)
autoignition point:	845 °F (452 °C)
ignition limits:	upper (not available)
	lower 2% (by volume)

While this is a typical ketone, there is much that is unknown as far as human physiological reaction to diethyl ketones is concerned. It is known to be hazardous when exposed to sufficient heat or flame, and to react quickly to oxidizing agents.

Extinguish diethyl ketone fires with dry chemicals, alcohol foam, or carbon dioxide.

Ethyl Acetone, $CH_3COC_3H_7$

Also known as pentanone-2 and methyl n-propyl ketone, ethyl acetone is a slightly water-soluble, colorless liquid with the following characteristics:

specific gravity: 0.8
boiling point: 215 °F (102 °C)
flash point: 45 °F (7 °C)
autoignition point: 940 °F (504 °C)
ignition limits: upper 8%
lower 2%   (by volume)

Alcohol foam, dry chemicals, or carbon dioxide should be used to fight ethyl acetone fires.

Other liquid ketones are:

Name	Specific Gravity	Boiling Point		Flash Point		Autoignition Point		Upper Ignition Limit (%)	Lower Ignition Limit (%)
		°F	°C	°F	°C	°F	°C		
ethyl acetone	0.8	215	102	45	7	940	504	8	2
cyclohexanone	0.9	312	156	110	43	788	420	–	–
diacetyl	1.0	190	88	80	27	–	–	–	–
phorone	0.9	387	197	185	85 (oc)	–	–	–	–
acetylacetone	1.0	282	139	105	41 (cc)	–	–	–	–

Extinguish fires with the above ketones with alcohol foam, dry chemicals, or carbon dioxide.

GLYCOLS

Glycols are, in a sense, double alcohols in that they have two hydroxyl groups (OH) on the chain. They are sometimes called dihydric for the same reason. In general, they have relatively high boiling points so that vaporization with sub-sequent inhalation is not usually a problem.

Ethylene Glycol, CH_2OHCH_2OH

Also known as glycol, glycolic alcohol, ethylene alcohol, and 1,2 ethanediol, ethylene glycol is a colorless liquid that is very soluble in water and has the following characteristics:

specific gravity: 1.1
boiling point: 388 ° (198 °C)
flash point: 232 °F (111 °C) (cc)
autoignition point: 775 °F (413 °C)
ignition limits: upper (not available)
lower 3% (by volume)

Ethylene glycol is known to react with strong oxidizing materials, but the danger of fire is normally small.

Fires may be extinguished with water, alcohol foam, dry chemicals, or carbon dioxide.

Propylene Glycol, $CH_3CHOHCH_2OH$

Propylene glycol, also known as methyl ethylene glycol and 1,2 propanediol, is a colorless liquid with the following characteristics:

specific gravity:	1.0
boiling point:	371 °F (188 °C)
flash point:	210 °F (99 °C)
autoignition point:	790 °F (421 °C)
ignition limits:	upper 13%
	lower 2% (by volume)

This is a relatively safe material, not given to self-heating and only a moderate fire hazard. It will, however, react with oxidizing agents.

Fire may be extinguished with alcohol foam or carbon dioxide.

SULFUROUS LIQUIDS

The element sulfur is a solid at normal temperatures and pressures, but because of its relatively mild chemical nature it can be kept as a liquid by simply sustaining its temperature above the melting point. Contrary to popular opinion, sulfur has no aroma as a chemically pure substance. It is any one of several common sulfur *compounds* that smells bad and gives sulfur its bad reputation.

Carbon Disulfide, CS_2

Sometimes called carbon bisulfide, carbon disulfide is normally a clear, colorless liquid that may turn a slight straw color with age. When absolutely pure it has no odor, but upon aging it can begin to smell somewhat like rotten cabbage. It has the following characteristics:

specific gravity:	1.3
boiling point:	116 °F (47 °C)
flash point:	−22 °F (−30 °C) (cc)
autoignition point:	212 °F (100 °C)
ignition limits:	upper 45%
	lower 1% (by volume)

Carbon disulfide is, by far, *the most hazardous liquid known.* It has all of the characteristics of a firefighter's nightmare; it is *highly* flammable, *highly* explosive, *highly* toxic, *highly* mobile, very inexpensive to produce, and very useful in industry, making it a *very common, very dangerous* liquid.

While carbon disulfide is not subject to spontaneous ignition, it is, nevertheless, an extreme fire hazard when exposed to just about any kind of energy input such as a spark, flame, heat, or friction. When exposed to such energy input, the chances of explosion are great if the air to carbon disulfide vapor ratio is in the vicinity of 95 to 5. It is an excellent candidate for a disaster-producing agent, because when it decomposes slowly it produces highly toxic gases, carbon monoxide and sulfur dioxide, thus:

$$2CS_2 + 5O_2 \longrightarrow 2CO + 4SO_2$$

When it decomposes rapidly, carbon disulfide will burn or explode as well as produce the toxic gases.

The low boiling point (116 °F, 47 °C) and high vapor density (2.2) of carbon disulfide assure not only considerable vapor production, even with the slightest rise in temperature, but also vapors that drift downward while not dispersing—almost as if they had intelligence and were looking for an ignition source. That ignition source is usually not too hard to find with an autoignition temperature of 212 °F (100 °C). The fact that the vapors are not water soluble makes it impossible to flush an area with fog nozzles.

The collective hazard potential of carbon disulfide dictates extreme care in storage and handling. Good housekeeping, good ventilation (with explosion-proof motors, if electrical), and good electrical grounding techniques are all absolutely required.

In fighting carbon disulfide fires, complete skin coverage and self-contained masks must be worn. While the specific gravity of 1.3 indicates that water can be used to smother its fire, extreme care must be used. Hydrogen sulfide, a highly toxic gas, can be generated in sulfide fires where water is used, so that putting out the fire may be more dangerous than letting it burn. If wind conditions allow it, water can be applied from an upwind position as long as the fire can be extinguished quickly and downwind population densities are properly protected. Blanketing with water, where applicable, has been successful when done with care so as not to disturb the surface. Both carbon dioxide and dry chemicals have been tried but without too much success. Water has been used with dry chemical using the water to cool hot surfaces preventing reignition. Where carbon dioxide has been successful, it has been poured onto relatively small fires in very large quantities. Foams are almost useless because the carbon disulfide vapors work their way through the foam structure and burn above it.

Amyl Mercaptan, $CH_3 CH_2 CH_2 CH_2 CH_2 SH$

Also known as 1-pentanethiol, amyl mercaptan is a usually colorless liquid that can be straw-colored to yellow. It has the following characteristics:

specific gravity:	0.8
boiling point:	260 °F (127 °C)
flash point:	65 °F (18 °C) (oc)
autoignition point:	not available

A true mercaptan, amyl mercaptan has an offensive odor and, what is more important, when it begins to decompose under the influence of heat it will develop very toxic vapors. Amyl mercaptan will react with oxidizers to generate more heat and flame as it begins to decompose. When water is sprayed on an amyl mercaptan fire, flammable gases will be produced as well as highly toxic vapors. It will react with steam in the same manner.

Amyl mercaptan fires should be extinguished with foam or dry chemical.

Dimethyl Sulfate, $(CH_3)_2SO_4$

Also known as methyl sulfate, dimethyl sulfate is a colorless liquid with the following characteristics:

specific gravity:	1.3
boiling point:	370 °F (188 °C)
flash point:	182 °F (83 °C)
autoignition point:	370 °F (188 °C)

The hazard of this compound is not so much from fire, which is only medium, but from any contact with the skin or mucous membranes of the body, which can result in delayed reactions, some of which have been reported as very serious. Death from liver damage has resulted even many weeks after the exposure.

Dimethyl sulfate fires may be extinguished by foam, dry chemicals, carbon dioxide or water.

Dimethyl Sulfide, $H_3C — S — CH_3$

Dimethyl sulfide, also known as methyl sulfide and 3-methylthiomethane, is a colorless liquid with a very unpleasant odor that has the following characteristics:

specific gravity:	0.8
boiling point:	99 °F (37 °C)
flash point:	below 0 °F (−18 °C)
autoignition point:	403 °F (206 °C)
ignition limits:	upper 20% lower 2% (by volume)

This is a very flammable substance when exposed to input energy in about any form. When dimethyl sulfide decomposes, it generates very toxic vapors along with highly explosive oxides of sulfur. It will react quite easily with oxidizers.

Dry chemicals or carbon dioxide have been used to fight dimethyl sulfide fires.

Sulfur Chloride, S_2Cl_2

Also called sulfur monochloride, sulfur chloride is a reddish yellow fuming liquid that has an oily appearance, a sharp, unpleasant odor, and the following characteristics:

specific gravity: 1.7
boiling point: 280 °F (138 °C)
flash point: 245 °F (118 °C)
autoignition point: 453 °F (234 °C)
ignition limits: not available

Sulfur chloride decomposes on contact with water or steam to form highly toxic vapors that may cause respiratory damage if exposure is sufficient. Concentrations of about 150 ppm have been reported as fatal to animals under test conditions for very short exposures.

Fires can be extinguished with dry chemical or carbon dioxide.

Thionyl Chloride, $SOCl_2$

Also called sulfurous oxychloride, thionyl chloride is a water white to straw-colored, and sometimes red, liquid with the following characteristics:

specific gravity: 1.6
boiling point: 174 °F (79 °C)
flash point: not available
autoignition point: not available
ignition limits: not available

Thionyl chloride will produce heavy fumes when exposed to air. These fumes smell somewhat like sulfur dioxide, and reacting even with the moisture in the air, they will produce the very toxic hydrogen chloride and sulfur dioxide compounds. When heated, thionyl chloride will decompose very rapidly. It will react vigorously with large quantities of water to produce disastrous quantities of these toxic vapors.

Sulfuryl Chloride, SO_2Cl_2

Also called sulfonyl chloride, sulfuric chloride, chlorosulfuric acid and sulfuric oxychloride, sulfuryl chloride is a water white liquid with a very disagreeable odor, and has the following characteristics:

specific gravity: 1.7
boiling point: 156 °F (69 °C)
flash point: not available
autoignition point: not available
ignition limits: not available

This is a typical sulfur compound that, when heated, will produce highly toxic decomposition products. It will also react vigorously with water or steam in an exothermic reaction to produce very corrosive and toxic vapors.

Although sulfuryl chloride is highly corrosive, it will not burn. It is, nevertheless, a very hazardous material.

ORGANOMETALLICS

With the blooming of the plastics industry, as well as antiknock research for the gasoline engine, there came on the scene a series of compounds that are purely synthetic—they are never found in nature. These are the organometallic compounds which are the chemical combination of a metal and one or more organic radicals. In these molecules the metal is, more often than not, the center of or the joining link between the several hydrocarbon chains or rings.

There are over 1000 of these compounds now known. About 300 are liquids, almost 700 are solids, and a few are unstable gases. Not all of these substances are useful; some are merely laboratory curiosities, but those that are useful find favor because of their high reactivity. This high reactivity is the very characteristic that also makes them a hazard. The more common liquid organometallics are, in general, extremely hazardous to handle; they burn with little or no initiating energy, react with air and, in some cases, violently with water, foam, and even the halon extinguishing agents. When burning, many of these compounds produce combustion products that are themselves combustible and difficult to extinguish. They therefore require great expertise and special personal protection when being handled or when fighting their fires.

There is one notable exception: tetraethyl lead. While this compound is flammable, it does not react with water or air spontaneously as do most of the others of this family. It is, nevertheless, a very toxic substance and responsible for much of the world's toxic pollution, having been decomposed in the almost infinite number of the world's gasoline engines.

A few of the more common organometallics and their characteristics are given in Table 6-1.

ORGANIC PEROXIDES

As described earlier, the inorganic peroxides tend to be unstable molecules with high chemical reactivity. The organic peroxides have, in addition to the peroxide linkage ($-O-O-$), hydrocarbon groups replacing hydrogen atoms. These become large, unstable molecules and are so reactive that they have been described as "disasters looking for a place to happen." They are very sensitive to shock and heat, and most of them, in their pure state, are so prone to spontaneous reaction that the Department of Transportation of the United States will not allow them to be shipped uninhibited.

All organic peroxides are irritating to the eyes, skin, and mucous membranes, and many produce toxic vapors if they are heated above storage temperatures. They are all excellent oxidizers and for this reason are used extensively in the plastics industry. Because they are such extremely good oxidizers, they must be kept well isolated from other organic materials, at cool temperatures, but not cold enough to freeze them, and in a storage area that is well ventilated.

Name	Formula	Color	Specific Gravity	Boiling Point	Decomposition Temperature	Air Reaction	Water Reaction	Fire Extinguishment
butyl lithium	–	straw	–	–	–	ignition	violent	dry chemicals
nickel carbonyl	$Ni(CO)_4$	colorless to yellow	1.3	109 °F 43 °C	140 °F 60 °C	explosion	–	dry chemicals
dimethyl cadmium	$(CH_3)_2\,Cd$	colorless	2.0	223 °F 106 °C	223 °F 106 °C	explosion	decomposes	dry chemicals
trimethyl aluminum	$(CH_3)_3\,Al$	colorless	–	266 °F 130 °C	300 °F 149 °C	ignition	detonation	dry chemicals dry sand
diethyl beryllium	$(C_2H_5)_2\,Be$	colorless	–	230 °F 110 °C	–	–	decomposes	dry chemicals
tri-isobutyl aluminum	$(C_4H_9)_3\,Al$	colorless	0.8	decomposition		ignites below 40 °F (4 °C)	violent	carbon dioxide dry chemicals dry sand
tetraethyl lead	$(C_2H_5)_4\,Pb$	colorless	1.7	392 °F 236 °C	260 °F 162 °C	–	–	water

Table 6-1.

Many solid organic peroxides may be found as liquids, or as what seem to be liquids, because they have been dissolved in a solvent pasted with an organic liquid, or in some cases water, in order to inhibit them. The liquid organic peroxides have low flash points, add to their overall danger by more easily generating toxic vapors than their solid counterparts, and if frozen become even more sensitive to shock than solid organic peroxides at normal temperatures. In addition, their inhibiting solvents, while reducing the chances of a runaway peroxide reaction, may contribute to the vapor hazard potential. A few of the more common liquid organic peroxides are discussed below.

t-Butyl Peracetate, $CH_3CO(O_2)C(CH_3)_3$

Also labeled tert-butyl peracetate, in industrial practice t-butyl peracetate is found as a liquid usually dissolved in benzene, mineral spirits, or some alcohols. Its characteristics vary over a wide range, depending on the solvent. However, safety dictates that the flash point be considered less than $80\,^\circ F$ ($27\,^\circ C$).

With benzene as the solvent, additional toxic, and possibly carcinogenic, vapors are added to its hazard.

Dry chemicals or carbon dioxide can be used in extinguishing fires.

t-Butyl Perbenzoate, $C_6H_5CO(O_2)C(CH_3)_3$

Also labeled tert-butyl perbenzoate, t-butyl perbenzoate is a clear to slightly yellow liquid often found in concentrated solutions at about 90 to 95 percent purity. It has the following characteristics:

specific gravity:	1.0 (and up)
boiling point:	$234\,^\circ F$ ($112\,^\circ C$)
flash point:	$190\,^\circ F$ ($88\,^\circ C$)
freezing point:	$46\,^\circ F$ ($8\,^\circ C$)

A typical organic peroxide, this material will cause combustion of almost any organic material it comes in contact with if not handled properly. It is very sensitive to shock and heat.

Fight fires with foam, dry chemicals, or carbon dioxide.

t-Butyl Peroxyisobutyrate, $C_8H_{16}O_3$

Also labeled tert-butyl peroxyisobutyrate, t-butyl peroxyisobutyrate is a clear to slightly yellow liquid usually found as a solution of the material in benzene, or in benzene with hydrocarbon solvents added. It has the following characteristics:

specific gravity:	0.9 (depends on solvent)
flash point:	at least $80\,^\circ F$ ($27\,^\circ C$)

The benzene solvent, when used, adds to the toxic hazard as a possible carcinogen. Care must be exercised with this material because the evaporation of the solvents causes increased sensitization to shock and heat. Age, container condition, and history of use are all factors to be taken into account when this material is involved in a fire.

t-Butyl Peroxypivalate, $(CH_3)_2 COOCOC(CH_3)_3$

Also known as tert-butyl peroxypivalate, the colorless liquid t-butyl peroxypivalate is usually found as a solution of an organic solvent and the material. It has the following characteristics:

specific gravity:	0.9 (varies with concentration)
boiling point:	violent decomposition above 40 °F (5 °C)
flash point:	above 155 °F (68 °C)

This peroxide cannot be heated above 30 ° to 40 °F (−1 ° to 5 °C) without exploding. It must be handled and shipped with refrigeration. Because it is such a good oxidizer, it is widely used in the polymer industry.

Use dry chemicals or carbon dioxide to fight fires.

Caprylyl Peroxide

Often found as a liquid, caprylyl peroxide is a solid in its pure form. For further details see the chapter on Hazardous Solids.

Cumene Hydroperoxide

Also called *a*-dimethyl benzyl hydroperoxide, cumene hydroperoxide is a colorless to slightly yellow liquid usually diluted with organic solvents so that its concentration is between 70 and 80 percent. In its pure form it has the following characteristics:

specific gravity:	1.0
boiling point:	307 °F (153 °C)
flash point:	175 °F (79 °C)

Cumene hydroperoxide, like all organic peroxides, is highly toxic and very irritating to the human body even when diluted to its normally found concentrations. It is a very strong oxidizing agent.

Attempt to extinguish fires with dry chemicals, foam, or carbon dioxide.

Methyl Ethyl Ketone Peroxide, $C_4 H_8 O_2$

Also known as MEK peroxide, methyl ethyl ketone peroxide is a commonly found liquid usually diluted to 30 to 50 percent in dimethyl phthalate. In concentrations above 60 percent it becomes extremely sensitive to shock and heat input.

Industrial solutions of this material exhibit a flash point of about 145 °F (63 °C). These diluted solutions will decompose with age, especially if exposed to temperatures above 250 °F (121 °C). They then become sensitive to the slightest disturbance. When heated rapidly to about 225 °F (107 °C) MEK peroxide will detonate.

This peroxide is a very dangerous, very toxic material, whose popularity makes it a common hazard.

Small fires involving MEK peroxide may be fought with great care using foam or carbon dioxide.

Peracetic Acid, $CH_3CO(O)_2H$

Also called peroxyacetic acid and acetyl hydroperoxide, peracetic acid is a clear liquid usually found as a mixture of the peroxide with acetic acid. It has the following characteristics as a 40 percent solution of peroxide in the acid:

specific gravity: 1.2
boiling point: 221 °F (105 °C)
flash point: 105 °F (40 °C) (oc)

This is a highly corrosive liquid that will severely burn living tissue. It has been known to explode with violence when heated slowly to 230 °F (110 °C). It tends to be very sensitive to shock at normal temperatures.

Carbon dioxide, foam, or water spray can be used to extinguish fires.

PETROLEUM PRODUCTS

Crude petroleum, as extracted from the earth, is a very complex, viscous mixture of liquids, gases, and solids. They are separated by a chemical process known as **fractional distillation**. This process makes use of the universal characteristics of liquids—different boiling points—by heating the crude mixture under the proper conditions and collecting the individual fractions as they boil off at their individual characteristic temperatures. The results of fractional distillation of crude petroleum are:

Fraction	Chain Length	Boiling Point Range	
		°F	°C
gases	C_1 to C_4	−265 to +86	−160 to 30
petroleum "ether"	C_5 to C_6	95 to 194	35 to 90
gasoline	C_7 to C_{10}	194 to 392	90 to 200
kerosine	C_7 to C_{10}	392 to 572	200 to 300
diesel oil	C_{11} to C_{16}	554 to 716	290 to 380
heavy oils	C_{20} and up	662 and up	350 and up
wax	C_{30}	melts	

Research on boiling point, vapor pressure, flash point, and other related characteristics of the fractions up to and including gasoline (C_{10}) shows that the shorter the chain length, the more physically unstable the liquid seems to be, and so, under the influence of the heat energy, breaks away from the pack to become a gas. Shorter chain length is related to greater volatility.

Of all the compounds that are distilled from crude petroleum, one, gasoline, is so common that it is almost impossible to get away from it in our present form of civilization. It may not be as commonplace as water but it must come close, and it is certainly far more hazardous.

Gasoline, C_7H_{16} (isomeric blend)

Gasoline that is refined from crude petroleum contains a great many different paraffin hydrocarbons from methane through decane, with many varied isomers. There are also a great many aromatics such as benzene, toluene, and isopropylbenzene in the mixture.

Gasoline, also known as petrol and just plain gas, is a clear liquid often colored with dyes to produce brand identification. It has a pleasant odor, is quite volatile, and has the following characteristics:

specific gravity:	variable but usually less than 1.0
boiling point:	100° to 425 °F (38° to 218 °C)
flash point:	−45 °F (−43 °C)
autoignition point:	495 °F (257 °C)
ignition limits:	upper 7.5% lower 1.4% (by volume)

The exact chemical formula for this mixture is quite variable depending on the producing company and even the general climatic conditions. The vapors themselves are not considered hazardous to life unless they displace, and therefore reduce, the normal life-supporting oxygen content of the area. The vapors are, of course, very dangerous when subjected to any energy input such as a spark, flame, or even sufficient heat. When enclosed, the vapors quickly reach the explosive limits and an explosion may result.

The ideal combustion process of gasoline may be written:

$$C_7H_{16} + 11O_2 \longrightarrow 8H_2O + 7CO_2 + \text{heat}$$

Under practical conditions, however, there is not always enough oxygen. Carbon and carbon monoxide will then be formed as the reaction runs fuel rich, with a sooty, yellow flame.

Fire extinguishment can be achieved with foam, dry chemicals, or carbon dioxide.

Natural Gasoline

Also known as casinghead gasoline, naturally occurring gasoline appears at the well head with the crude petroleum and other gases. It has approximately the same characteristics as the distilled product with the exception of its flash point, which is closer to, but still less than, 0 °F (−18 °C). Use foam to extinguish fires.

Kerosine, C_{10} to C_{16} (a methane series mixture)

Also called fuel oil number 1, coal oil, and range oil, kerosine is a clear to straw-colored, oily liquid with a strong aroma and the following characteristics:

specific gravity:	0.8 to 1.0
boiling point:	300 ° to 600 °F (149 ° to 316 °C)
flash point:	100 ° to 160 °F (38 ° to 71 °C)
autoignition point:	445 °F (229 °C)
ignition limits:	upper 5.0%
	lower 0.7% (by volume)

When the formula for kerosine is adjusted to keep the flash point above 100 °F (38 °C), as it should be, kerosine does not produce flammable vapors. When heated to the flash point, however, it is easily ignited, producing a considerable amount of heat. When kerosine burns in enclosed areas, as the available oxygen is reduced, more and more carbon monoxide is produced and less and less carbon dioxide. The total result can be deadly.

Kerosine vapors can explode when exposed to sufficient heat or an open flame. This does not occur too readily.

Extinguish fires with foam, dry chemicals, or carbon dioxide.

Benzene, C_6H_6

Also known as benzol, phenyl hydride and coal naphtha, benzene is a clear, water-white, pleasant-smelling liquid with the following characteristics:

specific gravity:	0.88
boiling point:	175 °F (79 °C)
flash point:	12 °F (−11 °C) (cc)
autoignition point:	1045 °F (563 °C)
ignition limits:	upper 7.0%
	lower 1.3% (by volume)

Benzene is a very common industrial substance whose vapors can, in high concentrations for even brief periods, cause serious injury to the nervous system. It is considered a carcinogen.

Benzene is a very dangerous substance in many ways. When exposed to sufficient energy input, it is a serious fire hazard and will explode when mixed with the proper volumes of air.

It is often used as the solvent in paints, lacquers, and varnishes, and may be found in gasoline, especially in the high octane grades.

Fires have been extinguished with foam, dry chemicals, and carbon dioxide.

Toluene, $C_6H_5CH_3$

Toluene, also known as toluol, methylbenzene, and phenylmethane, is a colorless liquid with the smell of benzene and the following characteristics:

specific gravity: 0.87
boiling point: 231 °F (111 °C) or higher
flash point: 40 °F (4 °C) (cc)
autoignition point: 950 °F (510 °C)
ignition limits: upper 7%   (by volume)
                 lower 1.3%

Nonlaboratory grades of toluene normally contain some benzene. Toluene is somewhat less toxic than benzene but it can, nevertheless, cause serious problems if inhaled in sufficient quantities.

Toluene is a highly flammable solvent used in a variety of industries. It will explode if the vapors are exposed to flame or spark and, when heated, produces toxic vapors. It reacts quite rapidly with oxidizing materials.

Toluene fires have been extinguished with foam, dry chemicals, and carbon dioxide.

Xylene (ortho, meta, and para), $C_6H_5(CH_3)_2$

Xylene, usually a mixture of the three isomers, is a water white liquid used as a solvent for a great many industrial applications, as well as an intermediate chemical compound in the formation of some plastics and medicines. The various characteristics are:

	Ortho		Meta		Para		Mixture	
specific gravity	0.880		0.864		0.861		0.864	
	°F	°C	°F	°C	°F	°C	°F	°C
boiling point	292	144	282	139	281	138	281	138
flash point	90	32	85	29	103	39	100	38 (cc)
autoignition point	865	463	980	527	985	529	950	510
ignition limits								
upper	6%		7%		7%		7%	(by volume)
lower	1%		1%		1%		1%	

Individually or as a mixture, the xylenes are a fire hazard when exposed to any form of energy input. They will explode when initiated and all will react with oxidizers.

During the combustion reaction, the xylenes produce considerable carbon and carbon monoxide, burning with a dirty, yellow flame.

Fires have been extinguished with foam, dry chemicals, and carbon dioxide.

Naphtha (mixture)

Also called coal-tar naphtha, benzol, hi-flash naphtha, Stoddard solvent, safety solvent and a variety of other names, naphtha is a water white to straw-colored liquid that can be a mixture of benzene, xylene, toluene, and many other petroleum-based or coal-tar based solvent liquids together or in selected combinations. The characteristics, therefore, will vary over a wide range about as follows:

specific gravity:	0.8 to 0.9
boiling point:	300 ° to 420 °F (149 ° to 216 °C)
flash point:	100 ° to 115 °F (38 ° to 46 °C) (cc)
autoignition point:	530 ° to 925 °F (277 ° to 496 °C)
ignition limits:	upper 6% to 7%
	lower 0.9% to 1% (by volume)

During the latter part of the nineteenth century, just about every new liquid that was discovered to be a good solvent seems to have been called naphtha. The names have stuck and confusion reigns.

While it is difficult to be specific about naphtha's characteristics, it is nevertheless true that its fire hazard is only medium if the vapors, which can be very explosive, are not allowed to collect. Naphtha can be very hazardous to the human body from either inhalation, ingestion, or even skin contact.

Fires have been extinguished with foam, dry chemicals, or carbon dioxide.

Fuel Oils

There are six different liquids in the classification of fuel oils, all with varying degrees of viscosity and from different fractions of the distillation process of crude petroleum.

The various characteristics are as follows:

Name	Specific Gravity	Boiling Point	Flash Point	Autoignition Point
fuel oil #1 (kerosine)	0.8 to 1.0	300 ° to 600 °F 149 ° to 316 °C	100 ° to 160 °F 38 ° to 71 °C	445 °F 229 °C
fuel oil #2 (diesel oil)	below 1.0		100 °F 38 °C	495 °F 257 °C
fuel oil #3	below 1.0		110 ° to 230 °F (cc) 43 ° to 110 °C	495 °F 257 °C
fuel oil #4	below 1.0		130 °F 54 °C	500 °F 260 °C
fuel oil #5	below 1.0		130 °F or higher 54 °C	
fuel oil #6	below 1.0		150 °F or higher 66 °C	765 °F 407 °C

Fuel oil number 3 through fuel oil number 6 are all very strong-smelling liquids with the resistance to flow increasing as the number of the fuel oil increases. Without exception, they are all only medium fire hazards and then only when exposed to high energy input. Fuel oils burn with black, dirty smoke and yellow flame. In spite of this, they can develop fires that can be difficult to extinguish.

Fires are usually extinguished with foam, dry chemicals, or carbon dioxide.

Petroleum Ether

Also known as benzine (note the "i"), petroleum spirits, ligroin, light ligroin, and petroleum naphtha, petroleum ether is *not* an ether, but really a mixture of straight chain hydrocarbons principally 5 and 6 carbons in length (pentane and hexane). It is a clear, water white liquid with a somewhat kerosine-like odor and the following characteristics:

specific gravity:	0.63 to 0.7
boiling point:	100 ° to 175 °F (38 ° to 79 °C)
flash point:	below 0 °F (−18 °C)
autoignition point:	550 °F (288 °C)
ignition limits:	upper 6%
	lower 1% (by volume)

This mixture is a very hazardous liquid when exposed to just about any form of energy input. While it will not flame spontaneously, it has been known to explode where confinement has allowed explosive vapor limits to be reached.

Sufficient inhalation of the vapors will produce an alcohol-like stupor, which proceeds from headache through coma.

Fires can be extinguished with foam, dry chemicals, or carbon dioxide.

A HALOGEN LIQUID

Bromine, Br_2

One of the members of the chemically active halogen family, bromine usually appears, at normal temperatures, as a dark, reddish brown, fuming liquid. It may also be seen in crystalline form in saturated solutions.

It has the following characteristics:

specific gravity:	3.0
boiling point:	138 °F (59 °C)
freezing point:	19 °F (−7 °C)
vapor density:	5.5

A noted research chemist once called this liquid "an evil genie in a bottle." It fumes its reddish brown vapors on contact with air, and because it is such a powerful oxidizer, bromine will initiate fire in combustible materials upon mere contact.

Like its family member, chlorine, it is a strong irritant to the eyes, nose, throat, and lungs. Fortunately, it is so strong an irritant that it usually provides ample warning for all but comatose personnel.

In an enveloping fire it generates highly toxic gases and will react with water in any form to produce very corrosive as well as toxic compounds.

In itself bromine is not a combustible material, but fires involving this material should be fought with carbon dioxide and *never with water.*

REVIEW QUESTIONS

1. What is evaporation? In light of the accepted molecular theory, explain what happens during evaporation.
2. Explain the state of equilibrium between liquid and gas states.
3. What is vapor tension? How is it measured?
4. Explain the mechanism of boiling. What is a normal boiling point?
5. How is boiling point influenced by pressure of the surface of the liquid?
6. What is flash point? How is it defined? How is it changed by dilution?
7. What is fire point? How does it differ from flash point?
8. What is autoignition point? How does it differ from flash point and fire point?
9. The following characteristics are useful in determining how a material reacts in a fire. Discuss what each tells us in a fire fighting situation: boiling point, flash point, fire point, autoignition point, vapor density.
10. What is viscosity? Explain how it produces different flow characteristics.
11. Discuss and compare the behavior of the following materials: aldehydes, alcohols, amines, ethers, ketones, glycols, sulfurous liquids, organo-metals, petroleum products.
12. When ethers stand exposed for a good length of time they form a very unstable compound. What is the mechanism of this formation and what are the potential hazards from this compound?
13. What are some of the important properties of acetone and how will these properties aid in fighting a fire?
14. Why do the physical properties of carbon disulfide make it so dangerous? Explain how it can be extinguished by blanketing.
15. What is gasoline and why is foam used to extinguish a fire in it? Why can't water be used on this fire?
16. Name the hazards associated with the aromatic hydrocarbons such as benzene and toluene. Why are their toxicities so insidious?
17. Discuss acrolein in terms of its physical properties and what they tell us about this material in a fire.
18. What are the vapor densities of toluene, ethyl alcohol, and pentane?
19. What is MEK? What type of liquid is it? Is it a hazardous material?
20. What is naphtha? Discuss the problems this presents for a firefighter.
21. Discuss the problems of organic peroxide fires.

Hazardous Solids/ 7

WHAT IS A SOLID?

A solid is defined as that state of matter in which the atoms are in a definite order producing a substance that has the ability to retain a definite shape, and possesses rigidity and volume. The ability to retain a definite shape and to possess rigidity and volume is due not only to the molecular activity but more to the lack of mobility that is the result of their interatomic or intermolecular relationships. These intermolecular and interatomic relationships may take one of two basic forms: crystalline or amorphous. The latter term means "without any definite shape," while the former term indicates that there is a very definite physical pattern to the arrangement of the atoms or molecules that make up the material. These patterns vary, with the result that the visible forms of the crystalline forms of the materials vary also. In addition, a single compound, or even an element, may have more than one visible, and therefore crystalline, form. These various forms of the same substance are called **allotropes**. They may possess different physical appearances and properties but nevertheless have the same chemical properties.

One of the very noticeable differences between crystalline and amorphous substances is the melting point. Crystalline materials have sharp, well-defined melting points; amorphous materials do not. The latter soften slowly and melt over a broad range of temperatures.

Three of the basic crystalline structures that are formed as a material solidifies are a cube (6 sides), a tetrahedron (4 sides), and an octahedron (8 sides). The crystals in these structures may arrange themselves in several different ways; a study of crystallography is beyond the scope of this book.

Because of the definite structure, it seems that there is a specific point at which the crystalline bonds begin to break away from each other, and the rigidity of the structure ceases to exist. The material then melts or liquefies. When there is no definite structure (amorphous), the rigidity disappears over a very broad range because there is no specific arrangement to the bonds.

While most solids will pass through the liquid state as their molecular energies are increased, and will then enter the gas state as the energy is further increased, there are a great many solids that pass directly from the solid state into the gaseous state. These materials, therefore, have no flash points. They do, however, have ignition points.

When, under normal room temperatures, a material passes from the solid to gas state without first becoming a liquid, that transformation is called **sublimation**. We may therefore say that a substance that undergoes this change but only under external, *abnormal* energy input, undergoes **pyrolytic sublimation**. This is a change of state, from solid to gas, caused by abnormal, external heat input.

This term is *not* to be used to describe a chemical reaction in which a solid disappears and a gas or gases are formed. Sublimation may only be indicated when the compound that was in the solid form is now the same compound as a gas. Wood at its ignition point is a good example of pyrolytic sublimation. As it approaches ignition, it becomes a gas, oxidizes, and so burns.

Note that pyrolytic sublimation is not a contradictory theory to that of the free radical mechanism for flame propagation. It is merely another, perhaps simplified, explanation of what happens during free radical production.

Nor is pyrolytic sublimation a substitute for ignition point. Ignition point is properly defined as the lowest temperature of the material at which a self-sustaining, exothermic chemical reaction will occur *for the existing conditions*.

That phrase, "for the existing conditions," is extremely important. The ignition point of a solid will vary with the moisture content of the material. Everyone knows that wet wood is more difficult to ignite than that which is dry. Vaporizing the water requires energy that is then not available to raise the temperature of the material to the point where pyrolytic sublimation begins and the resulting vapors will ignite.

Surface area does *not* influence the ignition point although it does influence the amount of energy needed to produce ignition. This is true simply because the mass of a small particle is so much less than that of a larger piece. With the smaller mass the temperature of that particle can be raised to its ignition temperature with much less energy. We all know that wood chips will ignite much faster than the log they once were. The ignition temperature, however, has not changed.

There is, in all cases, a time-temperature relationship in the ignition of solids. As the input temperature is increased, the time required for the material to generate vapors and reach its ignition point becomes less and less. When that time gets very close to zero time, the material seems to be exploding.

For solids with which we are most familiar, such as paper or wood, and which undergo pyrolytic sublimation, there is no melting point, boiling point or flash point. There are, however, many other solids that take the normal course of transition through the liquid phase and into the vapor phase. These solids have a melting point as the transition point between solid and liquid, and a boiling point as the transition point between liquid and gas. Once they become a liquid and as long as the temperature of a material remains high enough to keep it a liquid, it possesses hazardous liquid characteristics such as flash point and upper and lower ignition limits.

There are also solids that are so chemically reactive that it would appear it is not necessary for them to vaporize before undergoing combustion. These are known as **pyrophoric materials.** They react by simple exposure to the atmosphere. Actually a chemical reaction occurs with atmospheric oxygen at the surface of the material. This reaction is exothermic and produces flame.

The alkali metals are a good example of this class of solids. Pure sodium is said to have an autoignition point of 240 °F (116 °C), yet when exposed to room temperature (70 °F, 21 °C) air, it will ignite spontaneously. Actually the sodium reacts with the oxygen in the air to form an oxide:

$$4Na + O_2 \longrightarrow 2Na_2O + \text{heat and flame}$$

In contact with water, the reaction is even more violent because now a gas is produced:

$$2Na + 2H_2O \longrightarrow 2NaOH + H_2 + \text{heat and flame}$$

In this reaction, the sodium melts from the heat, and the hydrogen bursts into flame.

In both of the above examples the solid itself is not vaporizing but its unique chemical reactivity with another substance is producing a hazard. There are several metals in this pyrophoric class, as we shall see.

When discussing hazardous solids there is a danger of considering only those materials that are fuels. We must not forget that it takes two substances to produce a fire: a fuel and an oxidizer. There are many solids that are hazardous not because they burn or are toxic, but because they are the oxidizing agents.

CHARACTERISTICS

MELTING POINT

The melting point of a solid, a characteristic that is often used to identify a given material, is actually the threshold between that material existing as a solid and existing as a liquid. At the melting point both states will exist at the same time. This threshold is reached when sufficient energy is put into the molecules and their activity increases enough to overcome the forces that hold them in a rigid structure. If the input energy continues to be increased the entire substance will become a liquid. However, if, when the substance begins to melt, the amount of heat input is reduced so that the temperature of the material does not continue to rise, there will be an equilibrium set up so that for every molecule going into the liquid state, another molecule goes back into the solid state. The temperature at which this equilibrium between melting and freezing occurs is called the melting point of that substance.

It is quite possible, as a full liquid leaves that state to become a solid, that it may not cross the threshold between liquid and solid at the same temperature at which it went from a solid to a liquid. In such a case the substance remains as a liquid below its freezing point and is then described as **supercooled** or **undercooled.** Usually if a liquid in this condition is disturbed by a sudden mechanical impact or a single crystal of its solid is dropped into the supercooled liquid, the solid begins to form and heat is liberated as the temperature rises to the true melting/freezing point.

The contrails produced by high-flying aircraft are an example of supercooled water in the upper atmosphere being disturbed by the aircraft turbulence and thus going from a supercooled liquid (droplets) to a solid (ice crystals). It is these ice crystals that we see as a contrail.

If we study the X-ray patterns of solids, we find that while most materials in this state are in beautifully arrayed structures, there are some that have no order to their patterns, if in fact they can be called patterns at all. These substances therefore, meet part of the definition of a solid: they possess rigidity and volume, and have a definite shape. They do not have the ordered pattern of atoms. Are they, then, solids or liquids?

They are classified as supercooled liquids. Glass is an excellent example. Also considered in this class of solids are some plastics such as Lucite and Bakelite.

VAPOR PRESSURE OF SOLIDS

Solids, like liquids, can set up vapor pressure equilibrium in closed containers. Just as with any state of matter, the molecules or atoms have a range of energies. If at some instant in time, a unit near the surface exceeds the energy of those surrounding it, it can break free and become a gas molecule or atom. If the solid is in a closed container, it is possible for an equilibrium to be established as atoms or molecules break free of the solid and others return to the solid. As with liquids this equilibrium is not static, but a highly dynamic affair.

As it does with liquids, the vapor pressure of a solid is dependent on the temperature of the surroundings. The higher the temperature, the more energy will be available to the atoms or molecules and the greater will be the chance of their escape.

SOLUBILITY

While the concept of solubility is applicable to any combination of states of matter, we are most familiar with that of a solid dissolved in water. We see it every day as sugar dissolved in coffee or salt in cooking water.

The amount of any solid that will dissolve in a liquid is dependent on the nature of the liquid (solvent) and the nature of the solid (solute) and the temperature of the liquid.

Nature of the Liquid and Solid

The fact that a solid forms ions, and their number, greatly influences the type of solvent in which it will dissolve and to what extent. Water as a polar compound is an excellent solvent for strongly ionized solids, such as table salt, but is a very poor solvent for organic fats and other relatively nonionized solids.

However, nonpolar organic liquids, such as benzenes, are very poor solvents for ionized solids and are good solvents for the nonionized fats.

The Temperature

As we have seen so often, an increase in temperature causes an increase in molecular activity. This increase in activity manifests itself, in a solid, as more and more molecules possessing enough energy to break away from the solid and dash off into the liquid. For this reason then, *most* solids become more soluble in a given liquid as the temperature of that liquid rises. We must emphasize the word "most" because there are a few solids that become less soluble as the temperature rises. It appears, however, that this reduced solubility with increased temperature is due to the changes that take place in the intermolecular forces of the solid, which negate or work against the increased energy available to the individual molecules.

When all things are considered then, it must be remembered that an equilibrium point is *only* for *that* solid in *that* liquid at *that* temperature.

The quantity of solid that just sets up this equilibrium produces what is called a **saturated solution**. In the same manner, if the quantity of solid is *less* than that needed to produce equilibrium, the solution is said to be **unsaturated**.

Under normal conditions, once a solution reaches its saturation point, any additional solid that is put into the liquid will merely sink to the bottom of the container, or, if the solid is finely enough divided, will remain in suspension. In either case, it remains as a solid; it does not disappear into solution.

It is possible to dissolve more than the equilibrium quantity of a solid in a liquid but this is only accomplished under very specific conditions. A solution in this condition is known as a **supersaturated solution**. This condition may be achieved by heating the solution and dissolving in it the saturating amount of solid for that temperature and liquid. If the solution is then cooled very slowly and without any disturbance whatever, it is possible with many solids to have an abnormal amount of solid stay in solution. This can be done with sugar and water very easily.

These supersaturated solutions are delicate in that they are quite unstable. Just about any foreign substance or the slightest impact against the side of the container will cause the solid to kick out and the whole solution will become just another saturated solution.

The science of rain-making is built around the fact that many masses of air are supersaturated with water vapor, and are therefore just waiting for the disturbing conditions that will produce the kick out. The solid, silver iodide, has been found to be the best synthetic disturbance and is used by modern rain makers to produce water out of a few clouds. They seed the clouds with the silver iodide, and rain is the result.

IGNITION TEMPERATURE

Although solids do not oxidize to flammability directly but must first break down to free radicals or react to form a gas, they nevertheless are said to have an ignition temperature. This point is the lowest temperature at which the material will begin, and continue, a self-sustained exothermic combustion reaction.

When a solid does burst into flames, it forms a vapor that has about the same ignition point as the vaporization point. The solid then seems to flame at that point, but actually the solid, as such, is not burning. It is forming free radicals or complete molecules of gas. When we remove the heat given off by this reaction faster than it is being generated by the exothermic reaction, the fire will go out. When that exothermic reaction keeps producing free radicals or vapors the flames will keep rolling. Because it is an oxidation-reduction reaction, when we remove the source of oxygen or oxidizing source, the flames will then cease to exist. If, on the other hand, an oxidizer is burning, removing the fuel will produce the same results.

DELIQUESCENCE

All solids seem to have such an affinity for water that they will actually adsorb it from the surrounding atmosphere.

Those substances that are not water soluble (such as wood) will merely hold the water molecules with their physical structures, releasing them when the surrounding atmosphere drops in relative humidity. Some substances that are very soluble adsorb so much water vapor that they will actually form a liquid solution of the material and water. These materials are said to be **deliquescent**. Another word, little used by chemists but with about the same meaning, is **hygroscopic**. Some members of the scientific community would distinguish between the two terms by holding that solids that pick up moisture are deliquescent while liquids that do the same thing are hygroscopic. Under the above distinction, calcium chloride is a deliquescent compound and concentrated sulfuric acid is hygroscopic.

HYDRATES AND ANHYDROUS COMPOUNDS

Some compounds have such a great affinity for water that each molecule of the compound will physically bond itself to one or more water molecules. This is a direct combination but is not a chemical reaction, so that the water and the compound retain their identities and can be separated with relative ease. Because it is a physical attachment it is written as •H_2O. For example, potassium aluminum sulfate (alum) is written: $KAl(SO_4)_2 \cdot 12H_2O$. The alum is said to be **hydrated** with twelve molecules of water, which constitute the water of hydration and can be driven off by proper heating. When all of the water of hydration has been driven off, the material is then said to be **anhydrous** (without water). Only compounds that have hydrated forms have anhydrous forms.

EFFLORESCENCE

Because the stresses in the bond of hydration vary, some water of hydration requires less energy to be driven off than others. When the stress is very high the bond will break even at normal temperatures. When such a compound is exposed to an atmosphere of low relative humidity the water molecules will slowly break away until the hydrate becomes anhydrous or nearly so. This loss of water is called **efflorescence**.

SOME SPECIFIC SOLIDS

SULFUR

It wasn't without some foundation that the old masters painted hell as a place of fire fueled by sulfur, or brimstone as it was sometimes called. A sulfur fire has all that is fearful in the form of heavy, noxious yellow gases and hot, blue flames, together with puddles or perhaps rivers of molten sulfur.

In its pure state, and not burning, sulfur is a relatively nonhazardous solid. It is in combination with other elements (and it combines quite easily) that sulfur generates a serious hazard.

Because sulfur can exist in several different forms it has a unique property: the sulfur atoms tend to link together in an eight atom molecule generally in the form of a ring. When heated the molecular rings break up and relax into linear chains. This change in the structural relationship of the atoms causes the sulfur to melt at 239 °F (115 °C), but as the temperature of the melt, or liquid, is increased to about 390 °F (199 °C), the viscosity continually increases with an apparent increase in the thickness of the liquid. This is just opposite to the normal reaction of a liquid to a temperature increase. If the temperature of the liquid sulfur is increased above 390 °F (199 °C), the liquid seems to get thinner up to the boiling point (833 °F) (445 °C). Evidently the ring structures of the sulfur act like balls and flow over each other with greater ease than do the linear chains that form at the melting point. The chains, acting like fallen tree branches, become entangled with each other causing the thicker flow characteristics.

Elemental sulfur has the following characteristics:

density (solid):	2.07
melting point:	239 °F (115 °C)
boiling point:	833 °F (445 °C)
flash point:	405 °F (207 °C) (cc)
autoignition point:	450 °F (232 °C)

Pure sulfur is really a low hazard material, with the only potential hazard being that of its dust. This can be very real and serious. As with all dust situations, any spark or flame can result in an explosion.

In an enveloping fire, or when burning itself, sulfur will tend to expand the fire by flowing as burning rivulets. Sulfur fires produce quantities of toxic oxides that may endanger large areas.

To extinguish a sulfur fire, carefully use water but *only* as a fog; a water stream may well produce a steam explosion. Special dry chemicals may also be employed.

Polysulfides

If in the discussion of the structure of the sulfur atom, the sulfur rings and straight chains sounded like that of carbon chemistry, it is because there is a similarity. Sulfur atoms do bond together, as does carbon, but in nowhere near the number of combinations. These multiple sulfur combinations either link to each other in pure elemental sulfur or with a relatively few other atoms or groups to form what are known as polysulfides. For example:

Chemical Name	Common Name	Formula
iron disulfide	pyrite	FeS_2
ammonium polysulfide	—	$(NH_4)_2 S_x$
stannous sulfide	—	SnS_2
tetraphosphorous trisulfide	—	$P_4 S_3$
antimony trisulfide	antimony sulfide	SbS_3
antimony pentasulfide	antimony sulfide	SbS_5

Of the above compounds, the last two are perhaps the greatest hazards.

Antimony trisulfide is a red to black crystalline material with a density of 4.6. It can be a serious fire hazard especially in contact with strong organic acid, where it will react spontaneously. In contact with strong oxidizers such as the chlorates and perchlorates, it will react explosively. In an enveloping fire it will decompose to form sulfur dioxide and antimony dioxide, both toxic. When heated, or when burning, antimony trisulfide will react with water to produce very toxic and flammable gases. When cold it will react with steam to produce the same gases.

Fires may be extinguished with water, but only with the careful use of flooding or deluge techniques.

Antimony pentasulfide, having a density of 4.1, is an orange yellow powder that can be a serious fire hazard. This is especially true in an enveloping fire or in contact with strong oxidizing materials. It has been known to explode when impacted mechanically or in contact with strong inorganic acids. Like antimony trisulfide, it reacts with water or steam to produce very toxic and flammable gases.

Fires of antimony pentasulfide may be extinguished using water but only with the careful use of flooding and deluge techniques.

Sulfites

Replace the "d" in sulfide with a "t" and add three oxygen atoms to the sulfur atom; the result is another class of sulfur compounds called sulfites. Two examples are sodium sulfite (Na_2SO_3) and potassium acid sulfite ($KHSO_3$).

Sulfites are relatively unstable, because in contact with even weak oxidizing agents, they will take on another oxygen atom to become a sulfate.

As with most sulfur-bearing compounds, sulfites will, in an enveloping fire, decompose and produce sulfur dioxide (SO_2). This gas not only is very toxic but also will react with water in any form to produce corrosive sulfurous acid:

$$H_2O + SO_2 \longrightarrow H_2SO_3$$

Sulfates

When a fourth oxygen atom is attached to the sulfur atom, the compound becomes known as a sulfate. Examples include barium sulfate ($BaSO_4$), dehydrated calcium sulfate ($CaSO_4 \cdot 2H_2O$) (gypsum), monohydrated calcium sulfate ($CaSO_4$)$_2H_2O$ (plaster of paris), and sodium sulfate (Na_2SO_4).

In general the hazard potential of the sulfates is closely related to the element or elements to which the sulfate radical ($-SO_4$) is attached. Perhaps the best known of the sulfates is hydrogen sulfate better known as sulfuric acid (H_2SO_4)—a liquid.

Thiosulfates

There are many compounds where the affinity of sulfur atoms for each other seems to assert itself, and a sulfur atom takes the place of one of the oxygen atoms in the sulfate radical. This results in a radical that has multiple sulfur and multiple oxygen ($-S_2O_3$). For example, where sodium sulfate is Na_2SO_4, sodium thiosulfate is $Na_2S_2O_3$.

Except in enveloping fires, where they decompose to the toxic sulfur oxides, thiosulfates do not have a great hazard potential.

Persulfates

There is a class of compounds of sulfur in which the molecules are saturated with both sulfur and oxygen. These are called persulfates and in general are very strong oxidizers.

Potassium Persulfate, $K_2S_2O_8$

Also known as potassium peroxysulfate and anthion, potassium persulfate is a white crystalline solid with a density of 2.5. It is an excellent oxidizing agent that, when dry, will liberate oxygen just below 212 °F (100 °C). In solution it will release the oxygen at 60 ° to 70 ° (16 ° to 21 °C). In an enveloping fire, this material will promote burning as well as produce toxic sulfur oxides.

Ammonium Persulfate, $NH_4S_2O_8$

Also known as ammonium peroxysulfate, ammonium persulfate material is a white crystalline solid having a density of 2.0. It is considered a very strong oxidizing agent and should be handled with care when in the proximity of good reducing agents. Explosions have resulted from contact of the two classes. In an enveloping fire it will decompose at about 250 °F (121 °C) to produce toxic sulfur oxides.

Selenium and Tellurium

Two not-so-well-known members of the sulfur chemical family are selenium and tellurium. While elemental sulfur is a yellow solid, selenium is a red or gray solid and tellurium a silver gray solid.

Neither of these elements, in their pure state, has a high hazard potential. However, poisonings have been reported after long-term exposure to their dusts. The greatest hazard potential from these substances is from their compounds. This varies from zero to the very high toxicity of the organometallic compounds that contain these elements.

In an enveloping fire, the compounds of selenium and tellurium will decompose to produce highly toxic gases. In addition, as with any compound, high dust concentrations may be hazardous, especially where the potential of a spark input energy exists.

PHOSPHORUS

Elemental phosphorus appears in three distinct colors: white, deep red or violet, and black. They all have a wax-like appearance but differ greatly in their chemical reactivity and hazard potential.

Structurally, white phosphorus seems to be made up of many atoms held together by weak intermolecular forces, while the red phosphorus is several crystal lattices of atoms formed into layers. This layering seems to reduce the chemical reactivity by protecting the bonds of the interior atoms. White phosphorus, on the other hand, with its unprotected bonds and only weak forces holding the atoms together, is very active.

White phosphorus, which is sometimes called yellow phosphorus, has the following characteristics:

density: 1.82
melting point: 111 °F (44 °C)
boiling point: 535 °F (279 °C)
flash point: ambient temperature
autoignition point: 86 °F (30 °C)

When exposed to sunlight, this material can crystallize into the red form. However, it must be stored under water at all times, because when exposed to air it will immediately burst into flames. White phosphorus will react with explosive violence with most oxidizers, and so produce toxic gases that are essentially oxides of phosphorus. In fires where steam is present, these oxides can dissolve in the steam to form phosphorus and phosphoric acids:

$$P_4O_6 + 6H_2O \longrightarrow 4H_3PO_3 \qquad \text{phosphorous acid}$$
$$P_4O_{10} + 6H_2O \longrightarrow 4H_3PO_4 \qquad \text{phosphoric acid}$$

Inhalation of either of these acids can cause serious respiratory problems.

In nonfire situations, inhalation of the oxides of phosphorus will also produce respiratory problems when the oxides get into the lungs, react with the fluids there, and thus form the acids internally.

Fires are best extinguished with water using deluge techniques.

Red Phosphorus

Red phosphorus, which is also called amorphous phosphorus, is a violet red to brownish powder with the following characteristics:

density:	2.2
melting point:	1112 °F (600 °C)
boiling point:	sublimes
autoignition point:	500 °F (260 °C)

Although this form of phosphorus is considerably less reactive than the white form, and does *not* react explosively when exposed to air, it will, in an enveloping fire, produce toxic gases (those oxides again).

When red phosphorus is exposed to a prolonged heat input, it will react with oxidizers. It has also been known to explode when in contact with organic liquids of highly reactive properties.

Fires of this material are best flooded with water until the fire is extinguished and the area covered with wet sand.

Phosphorous Sesquisulfide, P_4S_3

Also known as tetraphosphorous trisulfide, phosphorous sesquisulfide is a yellowish gray solid with a crystalline form and the following characteristics:

density:	2.03
melting point:	344 °F (173 °C)
boiling point:	761 °F (405 °C)
autoignition point:	212 °F (100 °C)

As a combination of sulfur and phosphorus, this is a very toxic material. In an enveloping fire it will produce the oxides of both elements. In a water vapor situation or in contact with the interior surfaces of the lungs, some very toxic acids will be formed. Phosphorous trisulfide is reactive enough to explode in contact with most organic materials. It can also be ignited by an energy input in the form of simple friction.

Fires can be extinguished with water, as long as a flooding or deluge technique is used.

Phosphorous Pentasulfide, P_2S_5

Also known as diphosphorous pentasulfide, phosphorous pentasulfide is a greenish yellow crystalline solid with an odor like rotten eggs and the following characteristics:

density:	2.03
melting point:	529 °F (276 °C)
boiling point:	957 °F (514 °C)
ignition point:	288 °F (142 °C)
autoignition point:	549 °F (287 °C)

If exposed to air, this material will gradually pick up moisture and seem to melt; this is merely deliquescence, and is accompanied by the evolution of heat. The material can ignite if conditions are favorable. A reaction with the water will occur that generates hydrogen sulfide, a very toxic gas.

In an enveloping fire phosphorous pentasulfide acts like other phosphorus and sulfur compounds, forming the very toxic oxides of both. In addition, acids will be formed where water is available.

While fires have been extinguished with water, it should not be used except as a last resort. Carbon dioxide, or even sand, is better if available. Where water must be used, flooding or a deluge technique must be used along with self-contained masks for all personnel.

Phosphorous Pentachloride, PCl_5

Phosphorous pentachloride, a straw-colored to pale yellow crystalline solid, has the following characteristics:

density:	4.7
melting point:	333 °F (167 °C)
boiling point:	sublimes

When exposed to moist air, this solid will fume as it reacts with the water vapor. This is an exothermic reaction with enough heat output to result in a good fire. The reaction also generates very toxic chloric and phosphorous acids.

In an enveloping fire, phosphorous pentachloride will decompose to produce highly toxic chloride gases. These gases may then react with any available water to form very corrosive acids.

Attempt to extinguish phosphorous pentachloride fires with dry chemicals or carbon dioxide.

NITROGEN COMPOUNDS

It is not surprising, with all of the nitrogen present in the atmosphere, that there are so many nitrogen-based compounds available. What is surprising, with pure nitrogen so relatively inactive, is that so many of its compounds are very reactive even to the point of being explosive.

General Name	Radical Formula	Example
nitrides	$- N$	Na_3N sodium nitride
nitrites (inorganic)	$- NO_2$	$NaNO_2$ sodium nitrite
nitro- (organic)	$- NO_2$	$C_6H_5NO_2$ nitrobenzene
nitrates (inorganic)	$- NO_3$	KNO_3 potassium nitrate
nitriles (organic)	$- CN$	CH_2CHCN acrylonitrile
azides	$- N_3$	$Pb(N_3)_2$ lead azide
amides	$- \overset{\parallel}{\underset{O}{C}} - NH_2$	$HCONH_2$ formamide

Nitrides

The greatest hazard with most nitrides is that they will react with moisture in any form to produce ammonia. In an enveloping fire the ammonia which is thus developed has been known to react explosively in some cases, or at times merely to burst into flames.

Barium Nitride, Ba_3N_2

Barium nitride exists in the form of clear, colorless crystals having a melting point of 1832 °F (1000 °C). In contact with water, it has been known to undergo spontaneous oxidation. This reaction produces barium hydroxide and liberates flammable ammonia gas. When this ammonia oxidizes in air heated by the flames, more flame will result.

Fire fighters should attempt to extinguish such fires only with dry chemicals.

Lithium Nitride, Li_3N

Lithium nitride is a red brown crystalline material with a density of 1.3 and a melting point of 1553 °F (845 °C).

In contact with water in any form it reacts to produce the hydroxide (LiOH) and flammable ammonia gas. In an enveloping fire lithium nitride will burn in a normal atmosphere.

Fire should be extinguished only with dry chemicals.

Nitrites

While nitrites are not strong oxidizing agents, at least when compared to nitrates, they nevertheless have good hazard potential. Organic nitrites may, and, more often than not, do explode with violent decomposition. Nitrites are especially sensitive to shock leading to explosive decomposition. A mixture of a metallic nitrite and organic material provides an excellent invitation to an explosion.

In an enveloping fire, if nitrites do not explode they will decompose into toxic, gaseous oxides of nitrogen.

Ammonium Nitrite, NH_4NO_2

Ammonium nitrite, found as white to deep straw-colored crystals, is considered to be the most dangerous of the inorganic nitrites. The crystals have the following characteristics:

$$
\begin{array}{ll}
\text{density:} & 1.7 \\
\text{melting point:} & \text{decomposes} \\
\text{flash point:} & 158\,°F\ (70\,°C)
\end{array}
$$

Ammonium nitrite will react violently when impacted, shocked, or exposed to heat energy.

Fires are best extinguished with dry chemicals.

Potassium Nitrite, KNO_2

The straw-colored needles of potassium nitrite have the following characteristics:

$$
\begin{array}{ll}
\text{density:} & 1.9 \\
\text{melting point:} & 730\,°F\ (388\,°C) \\
\text{boiling point:} & \text{decomposes} \\
\text{ignition point:} & 1830\,°F\ (999\,°C)
\end{array}
$$

Potassium nitrite is considered a strong oxidizing material by the DOT. It is a medium fire hazard. When mixed with organic material, it can be ignited with very little input energy. Even the slightest friction has been known to set it off.

Fires should be extinguished with dry chemicals.

Sodium Nitrite, $NaNO_2$

Sodium nitrite, also known as diazotizing salts, may appear as either needles or a powder varying in color from white to straw. Sodium nitrite has the following characteristics:

$$
\begin{array}{ll}
\text{density:} & 2.17 \\
\text{melting point:} & 520\,°F\ (271\,°C) \\
\text{boiling point:} & \text{decomposes at about } 605\,°F\ (318\,°C)
\end{array}
$$

Like its family member, potassium nitrite, this material is also a strong oxidizing agent and a medium fire hazard. It too will explode at 1830 °F (999 °C) and should be kept from all contact with organic material.

Fight fires with copious amounts of water, using a flooding technique.

Nitrates

Nitrates are among the best oxidizing agents known. While they are considered only a medium fire hazard, they are generally serious explosive hazards either by way of spontaneous reactions or sensitivity to impact or shock. Many of them, such as ammonium nitrate, are in general use as fertilizers but can become high explosives when properly initiated.

In an enveloping fire, nitrates will decompose to produce highly toxic gases. Unlike their relatives, the metallic nitrates are not water-reactive. They will, however, adsorb moisture from the surrounding atmosphere and eventually dissolve (deliquesce).

Nitrate fires should be fought with water using deluge techniques, getting the water on as quickly as possible.

Ammonium Nitrate, NH_4NO_3

The clear, colorless crystals of ammonium nitrate have the following characteristics:

> density: 1.7
> melting point: 337 °F (169 °C)
> boiling point: decomposes at 410 °F (210 °C)

Ammonium nitrate is far more sensitive to energy input when it is impure. Just about any contaminant seems to sensitize it. If kept in a cool environment it becomes less reactive than if allowed to reach temperatures above 200 °F (93 °C). In these higher temperature situations, the amount of confinement is of prime importance. Above 200 °F (93 °C) the material may begin to decompose, even though very slowly, and during the decomposition heat is generating more heat and in a confined space increasing the pressure or at least insulating the system and so promoting further temperature increase. In either case the conditions promote reaction, pushing the rate faster and faster until it runs away as an explosion.

Fires should be extinguished by flooding with water and providing adequate ventilation.

Potassium Nitrate, KNO_3

Also known as nitre, saltpeter and niter, potassium nitrate is a clear to opaque finely crystallized powder with the following characteristics:

> density: 2.1
> melting point: 633 °F (334 °C)
> boiling point: decomposes at 752 °F (400 °C)

This excellent oxidizing agent is a medium fire hazard that becomes sensitized by mixing with just about any organic material. It then becomes so sensitive that the slightest input, such as friction, will bring about an explosion. In an enveloping fire it will explode if heated to about 1830 °F (999 °C).

Fires should be extinguished by flooding, or deluge, techniques with water.

Sodium Nitrate, NaNO₃

Also known as soda niter (or nitre), nitrate of soda and nitratine, sodium nitrate forms clear, colorless crystals that have the following characteristics:

$$\begin{array}{rl}
\text{density:} & 2.3 \\
\text{melting point:} & 585\,°\text{F} \ (307\,°\text{C}) \\
\text{boiling point:} & 715\,°\text{F} \ (379\,°\text{C})
\end{array}$$

Except for the above characteristics, sodium nitrate is very similar to potassium nitrate.

Fires involving sodium nitrate should be extinguished by water flooding, or deluge, techniques.

Nitro- Compounds

The "nitro" prefix is an addition to a compound name indicating an organic radical attached to an $—NO_2$ group. In general these compounds follow the nitrate characteristics of high flammability and sensitivity to explosion.

In an enveloping fire, they generally decompose to form highly toxic nitrogen oxides. If their rate of heat rise is rapid enough, they will explode.

AZIDES

Azides are primary explosives. They contain no oxygen and yet are among the best explosives known. When dry they are extremely sensitive to impact and shock as well as heat.

AMIDES

Amides are all organic, with the $—\overset{\displaystyle O}{\overset{\displaystyle \|}{C}}—NH_2$ group as the characteristic radical. They are not, of themselves, fire or explosion hazards. However, in combination with other groups they can become flammable and even primary explosives.

Amides, in their pure form, can produce toxic gases upon decomposition especially in an enveloping fire. In specific combinations with other hazardous materials they will, during decomposition, produce a vast range of toxic gases.

COMPOUNDS OF CHLORINE

Elemental chlorine is a gas at normal temperatures but many of its compounds are solids and some, liquids. As a combining atom chlorine exhibits many valence states. The more common ones are: $+1, +3, +4, +5,$ and $+7$.

Compounds of chlorine therefore exist as:

Valence	Name	Example	Formula
+1	chloride	sodium chloride	NaCl
+3	hypochlorite	sodium hypochlorite	NaClO
+4	perchlorate	sodium perchlorate	$NaClO_4$
+5	chlorite	sodium chlorite	$NaClO_2$
+7	chlorate	sodium chlorate	$NaClO_3$

It might appear from simple observation that the chlorine atom in the perchlorate grouping has a valence of +9. This is *not* the case. The perchlorate ion ($-ClO_4$) is a complex pyramid-like structure with oxygen atoms at each of the corners and the chlorine at the center. The valence bonds of the oxygen and chlorine are shared. Because of the pyramid structure this is a relatively stable atom and so perchlorates are relatively more stable than chlorates, chlorites, and chlorides.

Chlorides

As with other chemical compounds, the -ide suffix indicates no oxygen. Chlorides are therefore not usually flammable although in an enveloping fire, or in contact with inorganic acids, they can be very hazardous because they have the ability of producing highly toxic chlorine or reacting to form other chlorides. Organic chlorides such as carbon tetrachloride when heated produce the very toxic compound, phosgene.

Aluminum Chloride, $AlCl_3$

The straw-colored, coarse crystals of aluminum chloride have the following characteristics:

$$\begin{aligned}
\text{density:} \quad & 2.4 \\
\text{melting point:} \quad & 378\,°F\ (192\,°C) \\
\text{boiling point:} \quad & \text{sublimes at } 356\,°F\ (180\,°C)
\end{aligned}$$

Fire is not a great danger with this compound. However, in an enveloping fire, it will decompose to form with the moisture in the air highly corrosive gas. The reaction is exothermic and will produce considerable heat.

Fires should be extinguished by dry chemicals.

Some other chlorides are:

Name	Formula	Melting Point		Remarks
		°F	°C	
calcium chloride	$CaCl_2$	1422	772	nonhazardous below 2912 °F (1600 °C)
potassium chloride	KCl_2	1454	790	sublimes at 2732 °F (1500 °C); is nonhazardous.

Hypochlorites

With a valence of +3, the chlorine atom can pick up an oxygen atom (valence −2) and another electronegative element with a valence of −1 to change from a relatively inert to a very active compound. It becomes a strong oxidizer. These hypochlorites become active in an enveloping fire when the heat begins their decomposition into chlorine and oxygen.

In contact with an inorganic acid hypochlorites will react to form highly toxic chlorides and chlorine gas.

Sodium Hypochlorite, NaClO

Also known as the sodium salt of hypochlorous acid, sodium hypochlorite as NaClO does not exist except in water solutions. As a hygroscopic solid out of solution it exists as a colorless crystal with the general formula NaClO $2\frac{1}{2}H_2O$. It has the following characteristics:

> density: not available
> melting point: 136 °F (58 °C)
> boiling point: decomposes

Calcium Hypochlorite, CaCl(ClO) · $4H_2O$

Also known as bleaching powder, calcium hypochlorite is a white powdery material with the following characteristics:

> density: unstable
> melting point: decomposes
> boiling point: explodes if heated rapidly
> above 212 °F (100 °C)

In contact with just about any combustible material, calcium hypochlorite will decompose so rapidly the reaction rate might be called explosive. It is, in addition, a medium fire hazard at all times because of its highly reactive potential.

In an enveloping fire or in contact with inorganic acids, it will decompose to chlorine and oxygen. Both products need very little energy to go into explosive reactions. This compound will also react with water, especially steam, to produce corrosive acid and highly toxic gases. Contact with organic material can result in spontaneous ignition.

Chlorites

With a valence of +5, chlorine is able to unite with two atoms of oxygen (−2), as well as with another electronegative (−1) element such as an alkali metal. These resulting compounds are even more active than the hypochlorites.

Chlorites are used as commercial bleaching agents because of their excellent oxidizing characteristics. They will react explosively with sulfur.

Sodium Chlorite, $NaClO_2$

Sodium chlorite is a white, crystalline, deliquescent powder that does not seem to have a melting point but decomposes between 355 °F (179 °C) and 390 °F (199 °C) to produce chlorine and oxygen. The chlorine will unite with water vapor in the air to produce hydrochloric acid.

Sodium chlorite is very sensitive to impact or shock and has been ignited by friction alone. When impacted it will explode and in an enveloping fire will decompose to the very toxic gases. The oxygen present during decomposition may lead to explosions.

Fires involving this material should be fought from a full-stream distance with explosion protection.

Chlorates

The molecular structure of a chlorate must be very highly stressed because it is so unstable and sensitive to the slightest energy input. Chlorates are generally crystalline in form and tend to absorb moisture from the surrounding atmosphere.

Chlorates are very strong oxidizing agents and exhibit medium fire potential. This is especially so in the impure state where the impurities are oxidizable. When the impurities are organic or metallic in nature the compounds are prone to explosion. They are, in fact, used industrially as explosives.

Aluminum Chlorate, $Al(ClO_3)_3$

Aluminum chlorate compound is normally found as a hydrate with 6 molecules of water attached. It has been known to burst into flames spontaneously. This material is very sensitive to impact, heat, or mere contact with mild reducing agents. Aluminum chlorate will deliquesce if allowed to stand in air, and will decompose so easily that no melting point can be found.

When impure, this material can be very sensitive to all forms of energy input.

Small fires may be fought with dry chemicals, but large fires are so explosive they must be fought from full-stream distance with good explosion protection.

Potassium Chlorate, $KClO_3$

Also known as potassium oxymuriate and oxymuriate of potassium, potassium chlorate is a clear, white powder or colorless crystal with the following characteristics:

density:	2.3
melting point:	695 °F (368 °C)
boiling point:	decomposes above 752 °F (400 °C)

This material is a rather unstable compound that exhibits the general tendencies of all chlorates.

When mixed with charcoal and sulfur, potassium chlorate forms an explosive mixture, better known as black powder.

As with any chlorate, small fires may be fought with dry chemicals but large fires must be fought from full-stream distance with good explosion protection.

Other common chlorates are:

Name	Formula	Melting Point
ammonium chlorate	NH_4ClO_3	explodes at 215 °F (102 °C)
barium chlorate	$Ba(ClO_3)_2$	780 °F (416 °C) when dry
calcium chlorate	$CaClO_3$	212 °F (100 °C) when dry
magnesium chlorate	$Mg(ClO_3)_2$	95 °F (35 °C); decomposes above 248 °F (120 °C)

Fight all chlorate fires from a safe distance with plenty of water and explosion protection.

Perchlorates

While perchlorates are relatively stable, it should not be assumed that they cannot be hazardous. Perchlorates can be explosive or flammable compounds, especially in contact with organic compounds, carbon itself, some powdered metals, or sulfur.

All perchlorates are strong oxidizers because of the available oxygen that they carry in each molecule.

They are sensitive to impact or shock and in an enveloping fire produce very toxic gases.

A few of the more common perchlorates are:

Name	Formula	Melting Point
ammonium perchlorate	NH_4ClO_4	decomposes
barium perchlorate	$BaClO_4$	decomposes above 752 °F (400 °C)
potassium perchlorate (also known as potassium hyperchlorate)	$KClO_4$	1130 °F (610 °C)
sodium perchlorate	$NaClO_4$	decomposes above 896 °F (480 °C)
magnesium perchlorate	$Mg(ClO_4)_2$	decomposes above 482 °F (250 °C)
lithium perchlorate	$LiClO_4$	462 °F (239 °C); decomposes above 716 °F (380 °C)

Fight all perchlorate fires from full-stream distance, using as much explosion protection as possible.

HYDRIDES

The term hydride is derived from the word hydrogen and is used to indicate the presence of hydrogen with *no* oxygen. Aluminum hydride (AlH_3) is a good example. Hydrides are always the two element resultant product of a reaction between a metal and hydrogen. They may be divided into two classes, volatile and nonvolatile, depending upon which metal is in the compound.

In an enveloping fire, the more common nonvolatile hydrides, such as those of calcium and sodium, decompose to produce hydrogen. The volatile hydrides, such as those of phosphorous and boron, have been known to flame spontaneously in air. They too may decompose to produce hydrogen.

In short, then, the hydrides are a very dangerous group of compounds that may burst into flame or explode; they all produce hydrogen upon decomposition; and they will react with water in any form in an exothermic reaction to produce plenty of heat and hydrogen.

The hydrides also react violently with inorganic acids to produce heat and hydrogen.

with water: $AlH_3 + 3H_2O \longrightarrow 3H_2 + Al(OH)_3 + heat$

Sodium Hydride, NaH

An off-white to brownish powder, sodium hydride has all of the bad traits of its components sodium and hydrogen. A few of its characteristics are:

density: 0.9
melting point: decomposes at 1472 °F (800 °C)

By itself, sodium hydride is only a medium fire hazard. When it does begin to burn or is in an enveloping fire, however, it produces explosive hydrogen and the toxic oxides of sodium. In contact with any water it develops an exothermic reaction to produce heat, hydrogen, and sodium hydroxide, a very caustic liquid. In contact with oxidizing agents, it may react violently.

Attempt to extinguish sodium hydride fires only with dry powder, dry sand, or graphite.

Other common hydrides are:

Name	Formula	Density	Melting Point
calcium hydride (hydrolith)	CaH_2	1.8	decomposes at 1110 °F (599 °C)
aluminum hydride	AlH_3		decomposes
boron hydride (also known as decaborane)	$B_{10}H_{14}$	0.94	211 °F (99 °C)
lithium hydride	LiH	0.82	1256 °F (680 °C)
beryllium hydride	BeH_2		decomposes at 428 °F (220 °C)
magnesium hydride	MgH_2	1.4	decomposes above 392 °F (200 °C)

Fight any hydride fire only with dry powder, dry sand, or graphite.

CARBON

Carbon, with a melting point of 6458 °F (3580 °C) and a sublimation point at about 6600 °F (3649 °C), would seem to have a zero hazard potential. Such is not the case. Depending on its physical form, it can either burn and produce a great deal of energy; as dust, explode; or be practically inert. In one of its forms, bituminous coal, it is subject to spontaneous ignition.

The physical forms of carbon are:

Form	Common Name
amorphous	coal
8-sided crystals	diamond
6-sided crystals	graphite

Charcoal

This is a form of carbon plus impurities, whose physical appearance is that of a black, amorphous solid. It has been known to ignite spontaneously when wet, probably because of an exothermic reaction occurring in the impurities. Charcoal is one of the basic components of black powder, which is a mixture of potassium nitrate, sulfur, and charcoal. This mixture is classified as a Class A explosive.

Inorganic Compounds of Carbon

Although carbon is the basic element of all organic chemistry, there are nevertheless a great many inorganic compounds that contain this element. Even in these inorganic compounds, however, carbon maintains its valence of +4.

Like the compounds of nitrogen, sulfur, chlorine, and others, carbon compounds have been given suffixes to indicate the amount of oxygen present in the individual molecule.

For example:

Suffix	Example	Formula
-ide	calcium carbide	CaC_2
-ate	calcium carbonate	$CaCO_3$

Carbonyls

There is one other group of carbon compounds that exist in a gray area between inorganic and organic chemistry. These compounds are considered by some to be inorganic and by others to be organic, with the latter group seemingly in the majority. In any event, these compounds will, upon decomposition, produce organic substances. The group in question is called the carbonyls, which have a common radical $— (CO)_x$. For example, chromium carbonyl is $Cr(CO)_6$. Because most of the carbonyls are gases, we will discuss them at some length in the chapter on hazardous gases.

Carbides

Also known as acetylides, many of these substances are very dangerous. The metallic carbides are very sensitive to energy input of any kind. Impact, shock, friction, spark, or heat will often produce an explosion.

The alkali-metal carbides are not in themselves explosive, but will, on contact with water, generate an explosive gas. The metallic carbides are explosive in themselves.

Calcium Carbide, CaC_2

Calcium carbide is a gray, crystalline alkali metal carbide with a density of 2.2 and a melting point of 4172 °F (2300 °C). While the metallic carbides are safer when wet, the alkali metal carbides *must* be kept from moisture of any kind. In contact with water they produce acetylene, a potentially explosive gas. This reaction is:

$$CaC_2 + H_2O \longrightarrow C_2H_2 + Ca(OH)_2 + heat$$

Calcium carbide is considered a medium fire and explosive hazard but only because of the gas it generates on contact with water.

Carbide fires should be smothered with dry powder.

Copper Carbide, Cu_2C_2

Also known as cuprous carbide and cuprous acetylide, copper carbide is an amorphous reddish brown powder that does not melt when heated—*it explodes.* It is a very dangerous material!

Other more common carbides are:

Name	Formula	Density	Melting Point		Decomposition Point	
			°F	°C	°F	°C
aluminum carbide	Al_4C_3	2.4	2552	1400	above 3632	2000
boron carbide	B_4C	2.5	4442	2450	above 6332	3500
nickel carbide	Ni_3C	8.0	explodes			
lithium carbide	Li_2C_2	1.7	explodes			
sodium carbide	Na_2C_2	1.6	1292	700		

Examination of the formulas in the above table indicates why some carbides might be considered organic materials. In the simple valences of inorganic chemistry, sodium for example with its +1 valence would require Na_4C as its inorganic molecular formula. However, the molecular formula has been shown to be Na_2C_2 so the organic structure must be:

$$Na - C \equiv C - Na$$

This structure shows why the alkali metal carbides produce explosive acetylene (H— C ≡ C — H) so easily.

On the other hand, the metallic carbides seem to have inorganic structures. For example, nickel with its valence of +2 forms complex covalent structures with carbon. Its molecular formula has been shown to be Ni_3C.

Other Carbon Compounds
To specify all of the hazardous solid compounds of carbon is an almost impossible task. Nevertheless, some of the more common ones are listed below.

Naphthalene, $C_{10}H_8$
Also known as moth flakes, tar camphor, and white tar, naphthalene is a white, flake-form solid with a distinct odor and the following characteristics:

$$\begin{array}{rl}
\text{density:} & 1.15 \\
\text{melting point:} & 175\,°F\ (79\,°C) \\
\text{boiling point:} & \text{sublimes even at room temperature} \\
\text{flash point:} & 176\,°F\ (80\,°C)\ (cc) \\
\text{autoignition point:} & 980\,°F\ (527\,°C) \\
\text{ignition limits:} & \text{upper } 6.0\% \\
& \text{lower } 1.0\% \quad \text{(by volume)}
\end{array}$$

In spite of that relatively low flash point, naphthalene is considered only a medium explosion and fire hazard. In an enveloping fire the vapors can generate rapidly because of the low melting point. If confined, even in small pockets, the vapors will explode.

Fires should be extinguished with foam, dry chemicals, or carbon dioxide.

Camphor, $C_{10}H_{16}O$
Camphor, also known as gum camphor, 2-camphonone and lauryl camphor, is a white, crystalline material that has a strong distinctive odor and the following characteristics:

$$\begin{array}{rl}
\text{density:} & 1.0 \\
\text{melting point:} & 350\,°F\ (177\,°C) \\
\text{boiling point:} & \text{sublimes even at room temperature} \\
\text{flash point:} & 150\,°F\ (66\,°C)\ (cc) \\
\text{autoignition point:} & 870\,°F\ (466\,°C) \\
\text{ignition limits:} & \text{upper } 3.5\% \\
& \text{lower } 0.6\% \quad \text{(by volume)}
\end{array}$$

Camphor is only slightly less hazardous than naphthalene due to the higher melting point. Nevertheless, it does sublime slowly at room temperature and its vapors are explosive.

Fires should be extinguished with foam, dry chemicals, or carbon dioxide.

ORGANIC PEROXIDES

Solid organic peroxides have similar properties to the liquid organic peroxides in that they are all large unstable molecules, very sensitive to shock and heat, and usually, in their pure state, so liable to spontaneous chemical reaction that the Department of Transportation of the United States will not allow them to be shipped uninhibited.

Because they are such excellent oxidizers they must be kept in well-isolated storage at cool but not freezing temperatures. The area should be well ventilated. A few of the more common liquid organic peroxides follow.

Acetyl Peroxide, $(CH_3CO)_2O_2$

Also called diacetyl peroxide and ethanol peroxide, acetyl peroxide usually takes the form of colorless crystals. It has the following characteristics:

$$
\begin{aligned}
\text{density:} &\quad 1.2 \\
\text{melting point:} &\quad 86\,°F\ (30\,°C) \\
\text{boiling point:} &\quad 145\,°F\ (63\,°C)
\end{aligned}
$$

A true organic peroxide, acetyl peroxide is extremely sensitive to heat, shock, and just about any external energy input. Samples have been known to detonate spontaneously from aging only one day in a laboratory. Mere contact with other organic materials is likely to produce a combustion reaction. The material will react spontaneously if its temperature is raised above 86 °F (30 °C). Acetyl peroxide may also be found as a liquid when dissolved in the ratio of about four parts of solvent to one part of acetyl peroxide.

Fires should *not* be extinguished with water, since acetyl peroxide reacts with any form of water to produce more heat. Fight fires with dry chemicals or carbon dioxide.

Acetyl Benzoyl Peroxide, $C_6H_5CO(O)_2COCH_3$

Also known as acetozone, the white crystals of acetyl benzoyl peroxide may also be found in solution with the word "wet" or "solution" following the name. The "wet" usually indicates a paste of less than 40 percent of this peroxide in a solvent. The dry crystals have the following characteristics:

$$
\begin{aligned}
\text{melting point:} &\quad 97\,°F\ (36\,°C) \\
\text{boiling point:} &\quad 266\,°F\ (130\,°C)
\end{aligned}
$$

A typical, highly sensitive organic peroxide, acetyl benzoyl peroxide will detonate with the slightest provocation. An excellent oxidizer, it will react with water in any form to produce more heat.

Fire should be extinguished with dry chemicals or carbon dioxide.

Benzoyl Peroxide, $(C_6H_5CO)_2O_2$

Also known as lucidol, benzoyl peroxide is a white powder when dry but may be found wet in a paste form with water (about 30 percent peroxide by weight). If the water evaporates, the material becomes more and more dangerous as the water content drops. Benzoyl peroxide may also be pasted with other solvents.

It has the following characteristics:

melting point:	219 °F (104 °C) (decomposes)
boiling point:	explodes
autoignition point:	176 °F (80 °C) wet or dry

Benzoyl peroxide is a strong oxidizer that reacts with both organic and inorganic materials. It is very sensitive to external energy input and has been known to explode from being heated by the rays of the sun.

Fires may be extinguished with a heavy water spray or foam.

p-Chlorobenzoyl Peroxide, $(ClC_6H_4CO)_2O_2$

Also labeled Luperco BDB, p-chlorobenzoyl peroxide is, in pure form, a white powder. It is a very hazardous material that must be protected from all external energy input. There are many common materials that may act as a catalyst for its reactions, so it must be kept well isolated, even from dusts, in a cool, dry storage area. Laboratory explosions have been reported from mere contact with some organic dust.

Attempt to extinguish fires with foam or carbon dioxide.

Cyclohexanone Peroxide, $C_{12}H_{22}O_5$

Although cyclohexanone peroxide does exist as a powder, it is usually pasted with dibutyl phthalate (an oily liquid) in 40 to 80 percent concentrations of peroxide. At 85 percent concentration the material becomes extremely sensitive to shock and heat. While dibutyl phthalate is not very prone to evaporation at normal temperatures, aging of cyclohexanone peroxide pastes over long periods of time may cause extreme sensitivity to develop, and must be avoided.

Fires should be extinguished with dry chemicals, foam, or carbon dioxide.

2,4 Dichlorobenzoyl Peroxide, $(Cl_2C_6H_3CO)_2O_2$

Also known as Luperco CCC, 2,4 dichlorobenzoyl peroxide is usually found as paste made up of 30 to 40 percent solids in special organic solvents. At 50 percent concentration of the peroxide in a solvent, this peroxide becomes extremely sensitive to shock and heat.

Fires should be extinguished with dry chemicals, foam, or carbon dioxide.

Diisopropyl Peroxydicarbonate, $(CH_3)_2CHO(COO)_2CH(CH_3)_2$

Also called isopropyl percarbonate, IPP, and isopropyl peroxydicarbonate, diisopropyl peroxydicarbonate is a colorless crystalline material that will decompose rapidly above 52 °F (11 °C). It has the following additional characteristics:

density: 1.1
melting point: 48 °F (9 °C)

IPP is not only a powerful oxidizer but also a strong explosive. It must be kept refrigerated and not mixed with hot materials.

Small fires should be extinguished with dry chemicals, foam, or carbon dioxide.

Lauroyl Peroxide, $(C_{11}H_{23}CO)_2O_2$

Also known as Alperox C and dodecanoyl peroxide, lauroyl peroxide is a white granular powder that is a very strong oxidizing agent. It melts at about 126 °F (52 °C) and should be stored in a cool place with the temperature below 77 °F (25 °C).

Fires should be extinguished with dry chemicals, foam, or carbon dioxide.

Succinic Acid Peroxide, $(HOOCCH_2CH_2CO)_2O_2$

Also known as succinyl peroxide and butanedioic peroxide, succinic acid per-oxide is a white powder, usually found industrially about 95 percent pure.

Succinic peroxide has a melting point of about 255 °F (125 °C), at which temperature it begins to decompose. Because it is such a good oxidizer it is often found in the polymer industry.

Fires should be extinguished with dry chemicals or carbon dioxide.

Urea Peroxide, $CO(NH_2)_2 \cdot H_2O_2$

Also known as urea hydrogen peroxide, urea peroxide forms white crystals that have a melting point of about 175 °F (85 °C), at which temperature they begin to decompose. It is a typical peroxide with the added explosion potential of a nitrogen compound.

Fires should be extinguished with dry chemicals or carbon dioxide.

A HALOGEN SOLID

Iodine, I_2

One of the members of the halogen family, iodine usually appears at normal temperatures, as dark purples to black crystals with a somewhat metallic luster. It has the following characteristics:

density: 4.9
melting point: 235 °F (113 °C)
boiling point: 361 °F (183 °C)

Like its family members, chlorine and bromine, this element is a strong irritant to the eyes, nose, throat, and lungs. However, because it is a solid and does not vaporize very readily, it does not produce great quantities of vapor as do chlorine and bromine.

In an enveloping fire, iodine will generate highly toxic vapors and will react with water in any form to produce extremely toxic compounds. Such a fire should be extinguished with carbon dioxide, *never* water.

METALS

About 70 percent of all known elements are classified as metals because of their chemical and physical properties. Metals have the following characteristics:

Chemical
- In chemical reactions they tend to be electron donors.
- They form basic oxides.

Physical
- Metals conduct heat with efficiency.
- They conduct electricity with only reasonable loss.
- They have a metallic sheen on unoxidized surfaces.
- They exhibit good degree of malleability.

Not all metals will exhibit all of the above traits to the same degree. Some will excel in the conduction of electricity, others in malleability. They all have most of the characteristics in some degree.

In consideration of the potential hazards of metals, the first things that come to mind, at least for the more common metals, is that they are hard and can damage the body; they can be sharp and can cut the body; they retain heat energy and can burn the body.

However, over twenty metals can be radioactive and do considerable damage even though hidden. Over eighty metals can, under the proper conditions, oxidize rapidly enough to be considered flammable. The "proper conditions" vary with each chemical family. Some metals burn when exposed to water, while others burn only when dry; some burn only when exposed to air; others only when deprived of air.

The surface area of the metal is a very important factor in its ability to oxidize rapidly. In general, a smaller particle has a greater surface area, and therefore a greater tendency to react over a range from constant flame to explosion.

Chemically the metals may be divided into four groups: alkali metals, alkali earth metals, common metals, and noble metals.

The alkali and alkali earth metals are given these names because their oxides and hydroxides form very strong basic solutions with water.

The Alkali Metals

A few of the alkali metals and their properties are:

Name	Symbol	Density	Melting Point		Boiling Point		Common Radioactive Isotopes
			°F	°C	°F	°C	
lithium	Li	0.53	354	179	2435	1335	—
sodium	Na	0.97	208	98	1634	890	2
potassium	K	0.86	145	63	1382	750	2
rubidium	Rb	1.53	102	39	1292	700	2
cesium	Cs	1.9	84	29	1238	670	6

Note that in addition to their general metallic properties, the first three metals listed above have densities less than water and therefore float, and that the last three have very low melting points. One alkali metal, francium, has not been included because it is a laboratory curiosity rather than a material commonly found in nature or industry.

Fires involving alkali metals should be fought with dry powder.

Lithium, Li

In general, lithium and sodium have similar hazard potentials. However, sodium metal is far more plentiful and so presents a much greater hazard.

Lithium is a severe fire hazard especially in an enveloping fire or in contact with water in any form. It will react explosively on contact with inorganic acids or oxidizers. Under certain conditions lithium will react with nitrogen, which is normally considered a relatively inert gas. When lithium itself burns it produces toxic oxides.

Fires of the pure element should be extinguished with dry powder or special mixtures of graphite.

Sodium, Na

Sodium is a soft, putty-like material whose unoxidized surface has a metallic luster. In its pure stage, sodium is a very dangerous material that, even without burning but merely exposed to heat, produces highly toxic gases of sodium oxides. In an enveloping fire it will flame spontaneously in air and react vigorously with water in any form to produce explosive hydrogen.

Sodium will produce an exothermic reaction even with the moisture found on the surface of the skin. The result can be severe tissue burns both from the abstraction of the liquid and the development of sodium hydroxide, a very strong base:

$$Na + 2H_2O \longrightarrow 2NaOH + H_2 + heat$$

Fire control has been achieved with special dry powder mixtures, soda ash, absolutely dry sodium chloride, and graphite.

Sodium Solvent Mixtures

Also known as dispersed sodium and sodium dispersions, a sodium solvent mixture contains finely divided metallic sodium in an organic liquid such as naphtha, toluene, kerosine, or xylene.

These mixtures are extremely dangerous for two reasons: finely divided sodium is a hazard in itself, and the organic liquids are highly flammable. In an enveloping fire the solvents will ignite and burn exposing the raw sodium; this will then ignite producing a very persistent fire as well as highly toxic oxides in gaseous form. If water in any form is allowed to contact these mixtures when heated, the water decomposes, hydrogen is formed, and an explosion may well occur.

To extinguish a fire involving this mixture, special procedures must be used. Attempts may first be made to extinguish the sodium flames with special dry powder mixtures, and then the very careful use of carbon dioxide may be attempted to bring the organic liquid flames under control.

Potassium, K

Potassium is a silvery, lustrous crystal that is also extremely dangerous, reacting with water in any form to produce explosive hydrogen and potassium hydroxide as well as considerable heat. Quite often the heat output is so violent that it causes spattering of tiny spheres of molten potassium. This spattering will ignite the evolving hydrogen. If any confinement at all exists an explosion will result.

Even at normal temperatures potassium will react with atmospheric oxygen to form the peroxide (K_2O_2) and the superoxide (K_2O_4). It has been reported that potassium has extracted oxygen even from some oils under which it had been stored, and formed these same oxides.

Potassium peroxide and superoxide are yellow in color and very sensitive to the slightest friction. This is true even under oil.

In moist air potassium may develop enough heat to ignite spontaneously.

Fire fighters must attempt to extinguish such fires only with extreme care using great amounts of special dry powder mixes, soda ash, or graphite.

NaK

NaK, sometimes pronounced *nak*, is a unique mixture of potassium and sodium that has so low a melting point that it may appear as either a solid or a liquid. It combines the hazard potential of both elements, making it an extremely dangerous fire and explosive hazard. It reacts explosively with water or inorganic acids as well as with most oxidizers, and will flame upon contact with moisture in the air, burning vigorously and splattering molten drops of both elements in all directions.

Fires should be extinguished only with special dry powder mixtures.

Rubidium, Rb

Rubidium is a very soft, silver white, lustrous material that is dangerous in an enveloping fire where it produces toxic gases and oxides. Like other alkali metals, it will react with water to produce its hydroxide, hydrogen, and heat.

Rubidium will also react quite violently with chlorinated hydrocarbons.

Cesium, Cs

Cesium also is a silver white lustrous material with some ductility. It is a dangerous substance in contact with oxidizing materials or water in any form. Like the other alkali metals, in contact with water it produces its hydroxide, hydrogen, and heat.

Cesium will react quite violently with chlorinated hydrocarbons.

Alkali Earth Metals

The alkali earth metals are barium, beryllium, calcium, magnesium, radium, and strontium.

A few of their more important properties are:

Name	Symbol	Density	Melting Point		Boiling Point		Common Radioactive Isotopes
			°F	°C	°F	°C	
barium	Ba	3.6	1562	850	2795	1530	2
beryllium	Be	1.87	2336	1280	2732	1500	2
calcium	Ca	1.55	1562	850	2705	1485	2
magnesium	Mg	1.75	1204	651	2025	1107	—
radium	Ra	5.0	1760	960	2084	1140	4
strontium	Sr	2.6	1472	800	2489	1365	6

These alkali earth metals are harder than the alkali metals, ranging from barium which is about as hard as lead, to beryllium which will scratch most other metals.

Barium, Ba

Barium has a silvery white, metallic luster on unoxidized surfaces. With the exception of its radioactive isotopes, it is not considered a hazardous substance. In a finely divided state, however, its dust will ignite if subjected to flame and exposed in an enveloping fire. In this situation, with confinement, it has been known to explode.

Fire control has been achieved with special mixtures of dry powder.

Beryllium, Be

Sometimes called glucinium, this is a dirty white, lightweight but very hard metal. It is only a medium fire hazard and then only in a finely divided form. It will, in an enveloping fire or when heated by friction, burst into flame. Its dusts have been known to explode when exposed to flame in a confined area.

Fire control should be attempted with special mixtures of dry powder.

Calcium, Ca

Calcium is a white, soft metal that is considered a medium fire hazard. Like all of the alkali metals, it develops hydrogen on contact with water in any form. In confinement the hydrogen can produce an explosion.

Magnesium, Mg

Magnesium is a silver white, lustrous metal that can be very dangerous when the surface area is high. When in a solid block of reasonable size, magnesium does not ignite except at extraordinary temperatures. When the surface area is large, as it is with machine chips of this metal, magnesium can be ignited with the flame of an ordinary household match. Finely divided chips covered with cutting oil and water emulsion have been known to ignite spontaneously.

Magnesium will react with water in any form but especially with steam. In this reaction hydrogen is produced along with heat. The oxides produced as it burns are very toxic.

Fires should only be extinguished with special mixtures of dry powder.

Radium, Ra

The major potential hazard with radium is radiation. There are no common isotopes of radium that are not radioactive. All four isotopes are found in nature.

While it is not a fire hazard, should radium be involved in an explosion caused by other materials, the broadcast of its debris could be a major disaster.

Strontium, Sr

As with all other alkali earth metals, the greatest hazard potential with strontium is from its dust. It too will react with water to generate hydrogen and heat.

Commonly found as one of four artificial isotopes or two of its natural isotopes, the radiation hazard potential with strontium is great.

The Common Metals

There are many different ways of dividing the remaining metals into groups for ease of study. However, because we are not studying them as reacting chemicals but as hazards we shall group them together and only discuss the more common members.

Name	Formula	Density	Melting Point		Boiling Point		Number of Common Radioactive Isotopes
			°F	°C	°F	°C	
aluminum	Al	2.7	1220	660	4118	2270	—
antimony	Sb	6.7	1166	630	2507	1375	—
bismuth	Bi	9.8	520	271	2732	1500	4
cadmium	Cd	8.6	610	321	1409	765	3
chromium	Cr	7.2	2822	1550	3992	2200	1

Name	Formula	Density	Melting Point		Boiling Point		Number of Common Radioactive Isotopes
			°F	°C	°F	°C	
copper	Cu	8.9	1985	1085	4172	2300	1
iron	Fe	7.9	2795	1535	5072	2800	2
lead	Pb	11.3	622	328	2957	1625	3
manganese	Mn	7.2	2300	1260	3452	1900	3
molybdenum	Mo	10.2	4748	2620	10040	5560	1
thallium	Tl	11.8	576	302	3002	1650	4ˉ
tin	Sn	5.8	450	232	4091	2255	2
titanium	Ti	4.5	3137	1725	5432	3000	–
vanadium	V	5.9	3110	1710	⟩5435	⟩3002	1
zinc	Zn	7.1	788	420	1661	905	3
zirconium	Zr	6.5	3092	1700	⟩6335	⟩3502	3

Aluminum, Al

This silver white metal is essentially nontoxic, although like most solids it becomes a potential hazard as a fine powder.

When mixed with iron oxide, aluminum powder can be ignited and will develop temperatures high enough to melt iron. The mixture is known as thermite.

Control of fires involving aluminum must be attempted only with great care using special mixtures of dry powder.

Antimony, Sb

Originally called stibium, antimony is a silver gray metal which is very hard and brittle. While its compounds are very toxic, the element itself does not react at normal temperatures to form toxic gases.

In an enveloping fire, however, it will ignite if heated to about 800 °F (427 °C). In the form of dust antimony will explode. If held above its melting point, it will react with water to displace the hydrogen. At these temperatures and with the slightest confinement an explosion will result.

Fires should be extinguished with dry powder.

Bismuth, Bi

Bismuth is a silvery or reddish brown metallic crystalline material that can be very hard and brittle. It is a low hazard potential except for its radioisotopes.

It will ignite if held long enough in direct contact with a hot flame, and when burning it will produce toxic oxides.

Cadmium, Cd

Cadmium is a soft, silver white metal in its unoxidized state. However, because it reacts with air components, it is usually found as a black to reddish brown substance. Except for its radioactive isotopes cadmium is not a great hazard potential. Nevertheless, in an enveloping fire the high temperatures can cause toxic oxides to be generated.

Chromium, Cr

Chromium is a very hard, grayish crystalline metal that is essentially non-hazardous.

With the exception of its artificial isotope, ^{51}Cr, and the fact that as a dust it will ignite, it is a very stable substance. Such, however, is not the case with the compounds of chromium. Most of these are quite toxic and in some cases highly explosive—for example, chromium carbonyl.

Copper, Cu

The color of copper is so distinctive that it has become almost a generic color name for a reddish brown metallic hue. Except for the flammability of its dust at about 1300 °F (704 °C), it is not really very much of a hazard. Compounds of copper can be a different matter.

Iron, Fe

Once called ferrum, in its unoxidized condition iron is a silver gray, ductile metal that is not much of a hazard except in its dust form. The dust may ignite between 700 °F (371 °C) and 1400 °F (760 °C), depending upon the particle size, the atmosphere in which it is heated, and the amount of impurities in the metal. It has also been known to explode when its dust is exposed to heat and flame or in atmosphere with high oxygen content.

Lead, Pb

Once called plumbum, lead is a blue gray, relatively soft metal in its unoxidized state. It is a medium fire hazard in dust form with explosions having occurred between 1000 °F (538 °C) and 1300 °F (704 °C).

Where lead is involved in a fire, protective masks and clothing must be worn. Any lead or lead oxides that get into the human system by way of ingestion, inhalation, or even skin contact are not eliminated by natural functions, and will accumulate in body tissues. Nor does it take much accumulation to produce serious lead poisoning symptoms. Cases have been reported that have resulted from two or three exposures to smoke from fires enveloping lead products.

Manganese, Mn

Manganese is a reddish gray, brittle element that is a medium fire hazard but only when its dust or powder is exposed to direct flames. In this form it will ignite at about 850 °F (454 °C) and has been known to explode. This metal is the nearest common metal to the alkali earth metals in that it will react with water, especially steam, or acids in an exothermic reaction that produces explosive hydrogen. While the acid reaction will take place at normal temperatures, the water must be at 200 °F (93 °C) or higher to produce the reaction. Obviously this is not a difficult temperature to reach in a fire situation.

Enveloping fires involving manganese must be fought with full protective gear. Chronic manganese poisoning has occurred after a long history of even slight exposures to vapors and/or dust.

Fire control should only be attempted with special mixtures of dry powder.

Molybdenum, Mo

Molybdenum may appear as either a silvery metallic crystal or a dull gray black powder. It is only a slight hazard except when the particle sizes are very small. Dust or chips at about 1300 °F (704 °C) will produce steam if water is poured on them, and then react with the steam to produce hydrogen.

Fires should be extinguished only with special mixtures of dry powder.

Thallium, Tl

Thallium is a blue white, soft metal that is a medium fire hazard, but only in its dust form. While thallium is a very toxic material, it must be ingested to produce its full effects. It may be absorbed by way of the skin, but cases of poisoning by this method of entry are rare.

Thallium fires must *not* be fought with water, because it will form steam which then reacts with the thallium to produce hydrogen.

All four radioactive isotopes of thallium are artificial with half-lifes ranging from 3.7 years to 26 hours.

Tin, Sn

Once called stannum, tin can take on three forms, known as alpha, beta, and gamma tin. When exposed to extreme cold, tin may crumble into a gray, crystalline powder called alpha or gray tin. As a fine powder tin is only a slight fire hazard although a cloud of dust will ignite at about 1170 °F (632 °C). Tin will reduce water to oxygen and hydrogen at temperatures above 1300 °F (704 °C).

Titanium, Ti

Titanium may occur in either of two forms, a white, lustrous metal or a dark grayish black powder.

As small particles or dust it is a medium fire hazard when exposed directly to spark or flame. It will ignite at temperatures between 700 °F (371 °C) and 900 °F (482 °C). Dust clouds have been reported to explode at flame temperatures as low as 650 °F (343 °C). Large pieces of the metal have burst into flames at about 750 °F (399 °C).

In fires involving large quantities of titanium, water may be attempted as an extinguishing agent but only by a flooding technique. In small quantities, water will cause the formation of steam with the resultant production of hydrogen and a possible explosion. If possible, fires should be fought with special dry powder mixtures.

Vanadium, V
A dirty white to light gray metal, vanadium is not much of a hazard except for its dust which will ignite at about 900 °F (482 °C).

Zinc, Zn
Zinc, once known as spelter, is a light blue metallic, crystalline substance that, except for its radioisotopes and dust, is only a medium fire hazard.
    Fire control should be with special dry powder mixtures.

Zirconium, Zr
Also known as zirc, zirconium is a gray colored, lustrous metal or powder that is very dangerous both as a flammable substance and as an explosion hazard. When zirconium dust is exposed to direct flame it may burn, explode, or start to burn and then explode. With temperatures as low as 70 °F (21 °C), and with the dust well dispersed in air, it can be ignited very easily, and a violent explosion may result.
    Zirconium, as a finely ground powder, will explode more easily if it contains a slight amount of water (about 10 percent). It will even burn underwater once it is ignited, and will react with the water once the temperature of the metal is high enough. The water/zirconium reaction is as follows:

$$Zr + 2H_2O \longrightarrow ZrO_2 + 2H_2 + heat$$

Zirconium will also react with another normally good extinguishing agent, carbon dioxide, as follows:

$$Zr + CO_2 \longrightarrow ZrO_2 + C + heat$$

About 0.016 ounces of dust per cubic foot of air is considered to be a critical explosive content. Because of this and the fact that it will ignite at such low temperatures as 70 °F (21 °C), zirconium is known as a pyrophoric material.
    Fire control should only be attempted with special dry powder mixtures.

Alkali Metal—Common Metal Hybrids
There are some three-way combinations of alkali metals with common metals and oxygen that produce very hazardous compounds. These compounds are generally very strong oxidizing agents that tend to react violently with combustible materials of any kind. Two of the more common of these hybrids are sodium permanganate and potassium permanganate.

Sodium Permanganate, $NaMnO_4$
The deep reddish purple, almost black crystals of sodium permanganate are well known among chemists as excellent oxidizing agents. This hybrid decomposes in

an enveloping fire to produce oxygen and some very toxic oxides of sodium. It is, therefore, a very dangerous material and must be stored and kept away from all combustibles.

Potassium Permanganate, $KMnO_4$

Potassium permanganate forms deep purple crystals with a metallic luster, which have a density of 2.7. This compound is a very strong oxidizing agent that has been known to explode spontaneously upon contact with organic materials.

In an enveloping fire potassium permanganate decomposes at 415 °F (213 °C), producing toxic oxides of potassium and manganese.

REVIEW QUESTIONS

1. Define a solid. How does it differ from a liquid or a gas?
2. What is sublimation?
3. Write a paragraph on how surface area influences the ignition point and the amount of energy needed for ignition.
4. What are pyrophoric materials? List several.
5. Define melting point. Explain the action.
6. What is a supercooled liquid? How does it differ from an undercooled liquid?
7. Can solids have a vapor pressure? Explain.
8. What factors influence the solubility of a solid in a liquid? Explain how each influences the solubility.
9. What is an undersaturated solution? What is a supersaturated solution?
10. Explain deliquescence. How does it differ from hygroscopic?
11. What is the difference between hydrated and anhydrous compounds?
12. What is efflorescence? How does it differ from deliquescence?
13. Discuss the differences between polysulfides, persulfides, sulfates, sulfites, and thiosulfates.
14. Discuss the characteristics and activity of nitrogen and nitrogen compounds.
15. Define a metal. What physical and chemical characteristics make metals different from nonmetals?
16. Discuss the major problems of organic peroxides in an enveloping fire.

Hazardous Gases/ 8

WHAT IS A GAS?

A gas is that state of matter in which the material has no shape or volume of its own but will always assume the shape and volume of its container.

Even in the case of a very heavy gas, where one might expect to find it settled at the bottom, we always find some of the lighter molecules of that gas at the top of the container. The reverse holds true for a very light gas: instead of the total amount collected at the top, some of its heavier molecules would be at the bottom. This act of filling the container is called **diffusion** and is one of three important characteristics that can make a gas a potential hazard.

The three characteristics are: gases are extremely compressible; gases are infinitely expandable; and gases diffuse with ease.

The gas laws, as seen in Chapter 2, describe these characteristics with mathematical certainty. However, for fire safety personnel, these characteristics mean that a hazardous gas may be under great pressure, that if it is somehow released it will expand to fill any given area (if it doesn't explode first), and that it will mix readily with any other gases, including those which make up our atmosphere.

CHARACTERISTICS

COMPRESSIBILITY HAZARDS

According to the United States Government, Department of Transportation, Hazardous Materials Regulations, subpart G, section 173.300, a compressed gas is any substance or mixture of substances that develops an absolute pressure greater than 40 psi (276 kPa) at 70 °F (21 °C), or greater than 104 psi (717 kPa) at 130 °F (54 °C), or both. Because with enough pressure, and at the proper temperature, a gas can be compressed into the liquid state, the DOT also includes any flammable liquid that has a vapor pressure greater than 40 psi (276 kPa) at 100 °F (38 °C) in the compressed gas category, using a procedure detailed by the American Society for Testing Materials (ASTM).

The compression of a gas until it becomes a liquid is a common method of reducing the volume needed to ship many gases. When a gas is compressed the molecules are forced closer and closer together. Under proper conditions the intermolecular forces, called van der Waal's forces, take over, and at a very specific point the gas becomes a liquid. For each individual gas there is a temperature above which it *cannot* be liquified no matter how much it is compressed. This temperature is called the **critical temperature**, and is directly related to the amount of attractive, or van der Waal's, force the gas molecules of that compound have for each other. If the temperature is above the critical point, the molecules will possess too much kinetic energy and will resist the van der Waal's forces no matter how much compression is used to force them together. The gas does not become a liquid because the molecules never lock together. When the kinetic energy of the molecules is reduced, by removing the heat and lowering the temperature, to a point where the attractive forces can prevail, then, when the pressure forces the molecules close enough together, they will lock together and become a liquid. The pressure which must be applied at the critical temperature to bring about liquification is called the **critical pressure**.

A few relationships of critical temperature and critical pressure are:

Gas	Critical Temperature		Critical Pressure	
	°F	°C	(psi)	(MPa)
ammonia	270	132	1612	11.3
fluorine	200	93	807	5.6
hydrogen	−400	−240	185	1.3
oxygen	−182	−119	718	5.0
propane	206	97	617	4.3
nitrogen	−233	−147	484	3.4
methane	−116	−82	658	4.6
steam	705	374	3200	22.3

While a gas cannot be made to liquefy at temperatures *above* the critical temperature, it will liquefy at any temperature *below* its critical temperature as long as sufficient pressure is applied. The sufficient pressure becomes less and less as the temperature of the gas is lowered, the kinetic energy of the molecules is reduced, and the van der Waal's forces, which remain the same no matter what the temperature and the pressure are, finally take over to lock the molecules together as a liquid.

It also follows that should that pressure be released, the van der Waal's forces can no longer hold the molecules together and the liquid will become a gas. If the release is sudden, the rate of change from liquid to gas will be very rapid.

This transition from liquid to gas appears as a boiling action with considerable vapor being produced. Because the pressure has been suddenly released, there is no longer any confinement and the vapors expand rapidly. If the gas is flammable and the reason for the sudden release of pressure was a tank rupture in an enveloping fire, the expanding vapors, mixed with air, will explode. The result can be devastating, with a fireball several hundred feet in diameter reaching hundreds of feet into the air.

Because this explosion and resulting fireball are not unique phenomena, and because they are so spectacular when seen at a safe distance, a word has been coined to describe this situation: **BLEVE** (rhymes with "levee"), an acronym for the phrase *b*oiling *l*iquid *e*xpanding *v*apor *e*xplosion.

In the rapid expansion just prior to the explosion there is a substantial cooling of the liquid brought about by the sudden change of state from liquid to gas. It is not uncommon for a brief rain of cold liquid droplets to occur just as the explosion takes place. These droplets are thrown above the liquid by the violence of the boiling and are then propelled upward and broadcast by the explosion. They have been known to travel up to 2000 feet (610 metres).

A BLEVE will propel pieces of tank as well as droplets of liquid. Tank fragments have been thrown up to 1500 feet (458 m) and whole tanks 300 to 800 feet (91 to 244 m) depending on the size and weight of that container.

CRYOGENICS

We have seen that above the critical temperature no amount of pressure will liquefy a gas. We have also noted that the further below the critical temperature we reduce the temperature of a gas, the less pressure is needed to form the liquid. If we carry this idea to extremely low temperatures it should follow that every gas can be liquefied and many even solidified.

Cryogenic procedures do just that. Cryogenics is the study and use of substances at extremely low temperatures. In order to define what is meant by low temperatures, scientists have decided that $-150\,°F$ $(-101\,°C)$ is the cutoff point. Gases with critical temperatures below $-150\,°F$ $(-101\,°C)$ are often called **cryogens.**

Cryogenic Hazards

The hazards of cryogens, or any gas at cryogenic temperatures, are:

- the hazards specific to that gas whether at cryogenic or normal temperatures
- the extremely great compression which has taken place resulting in high liquid to gas ratios
- the extremely low temperature of the liquid mass or masses, which may destroy human tissue, liquefy other gasses, and solidify other liquids.

The hazards specific to a gas, whether at cryogenic or normal temperatures, are less with that material as a liquid than as a gas. As with any comparison between the chemical reactivity of a gas and a liquid, the liquid exhibits less activity than the gas; there is less surface area upon which the reaction can take place. However, chemical activity does not disappear or change with respect to the materials with which that material will normally undergo chemical reactions.

Liquid oxygen is chemically stable and essentially not flammable. However, it is still an excellent oxidizer, and, when sprayed into a rocket motor, unites with the fuel in a roaring reaction. The spraying does no more than promote the reaction by increasing the surface area.

Liquid to Gas Ratio

The liquid to gas ratio is an extremely important consideration for fire safety personnel. Cryogenic gases do not stay liquefied unless they are kept below their critical temperatures. Because these temperatures are abnormal, special designs of double walled, vented containers are used with a partial vacuum between the walls. However, in larger storage systems, a refrigeration mechanism must be kept in operation to counteract temperature rises and to keep the pressure within the container or tank from rising above the failure point of the material. Consider that liquid oxygen has an expansion ratio of over 860 to 1 and liquid fluorine almost 1000 to 1 between their critical temperatures and 70 °F (21 °C). Imagine the pressures that can develop in an enveloping fire where the temperature of the tanks will go far above 70 °F (21 °C) with a refrigeration mechanism that was never designed to handle such rates of temperature increase.

Low Temperature Mass

It is obvious that a cryogenic liquid is a concentrated form of the gas. As such its density is much higher, and so it has a greater potential for the removal of heat from any object it may contact. When cryogenic liquids contact human tissue they kill it by freezing the liquids and destroying the living cells. In controlled situations, skilled dermatologists may use liquid nitrogen to remove warts and moles from a patient's skin. Just a bit of the liquid on a swab is all that is needed to destroy the errant tissue.

In order to prevent uncontrolled contact of the cryogen with living tissue, every precaution must be taken to protect the face, eyes, and skin when in proximity to cryogens. Special clothing designed for these situations is recommended.

The mass and low temperature of the cryogen will also liquefy other gases It is not uncommon to see air liquefy and solidify when it comes in contact with other cryogenic liquids.

Liquid hydrogen has a boiling point of −423 °F (−253 °C). This temperature is so low that it has been known to solidify free oxygen. When in such a solid mixture, and in a confined area, the combination has resulted in serious explosions.

TYPES OF GASES

Gases may be divided into two categories: those that are chemically active and those that are inert.

Inert gases are those that do not, under normal circumstances, react chemically with other elements. They are:

Name	Symbol
helium	He
neon	Ne
argon	A
krypton	Kr
xenon	Xe
radon	Rn

The above gases are all colorless substances that neither gain nor lose electrons and so do not combine with other elements. They are not hazardous except that they may displace air from a confined area and thus deprive the body of life-supporting oxygen.

Active gases may be further divided into those that are chemically active outside the body and those that are chemically active in the body. The latter are toxic.

The active external gases may be categorized as:

1. those that react to produce heat
2. those that react to produce both heat and flame
3. those that react to produce toxic gases

The active gases range in chemical activity from just barely to very reactive. The activity may vary from internal exothermic activity to reactions that will proceed only under the influence of considerable heat or pressure (or both).

Flammable gases are those that react with atmospheric oxygen to produce visible heat output. These gases have a specific mixture range of their vapors with air that will, when ignited, produce flame and products of combustion.

Nonflammable gases are those that will not react with atmospheric oxygen no matter what the vapor to air ratio. It is possible to have certain specific mixtures of nonflammable gases that, under exacting and proper conditions, will support a redox reaction rapidly enough to produce flames. Oxygen-helium mixtures, used in some special welding techniques, are one such combination. These reactions may also produce toxic materials.

Chemically active gases are those that readily react with many other elements to produce heat, flame, toxic materials, or all three. The halogen family, especially chlorine and fluorine, are good examples of this group.

Some organic gases are so reactive that they will reorient themselves in response to certain energy inputs such as shock or heat, resulting in very hazardous materials. Acetylene and vinyl chloride are two such gases. They are normally shipped with inhibiting materials to reduce this sensitivity.

Toxic gases are those that of themselves present hazards to life or property when they are allowed to expand into the life support area. These gases may attack the outer skin or inner membranes and react chemically with the substances there. The products of these reactions or the removal of vital body chemicals jeopardizes the continued existence of the body.

Pressurized gases are those that have been packed for handling by compressing them into suitable containers. While this packing makes for economic shipping, it develops potential hazards because of the pressures involved. Even inert gases can be hazardous under these conditions.

SOME SPECIFIC GASES

AIR

Not normally considered a hazardous material, air can be dangerous under certain conditions. Its composition when dry is approximately:

Gas	Formula	Percent by Volume
nitrogen	N_2	78.09
oxygen	O_2	20.95
argon	A	0.93
carbon dioxide	CO_2	0.03
neon	Ne	0.002
helium	He	0.0005
methane	CH_4	0.0002
hydrogen	H_2	trace
ozone	O_3	trace
nitrous oxide	NO_2	trace
xenon	Xe	trace

Air has no chemical properties of its own; it is a mixture, in which the individual gases react in their own characteristic manners. However, because two chemically active gases, nitrogen and oxygen, comprise 99.04 percent of the mixture, air takes on the characteristics of these gases.

Air has the following characteristics:

vapor density: 1.0
boiling point: $-308\,^{\circ}F\ (-189\,^{\circ}C)$
flash point: none
autoignition point: none

Compressed air is stored in *green* labeled cylinders.

OXYGEN, O₂

That friendly enemy, oxygen, is normally a life-giving gas, but it may also be a hazardous liquid. Below its critical temperature of $-290\,°F\,(-179\,°C)$, it is often stored as a liquid and bled off for use, sometimes as a gas, sometimes as a liquid.

Oxygen is colorless and odorless as a gas, but as a liquid may have a slight bluish color to it. As a liquid it is usually referred to as lox, short for liquid oxygen.

It has the following characteristics:

$$
\begin{array}{rl}
\text{vapor density:} & 1.43 \\
\text{boiling point:} & -297\,°F\,(-183\,°C) \\
\text{autoignition point:} & \text{nonflammable} \\
\text{explosive limits:} & \text{none}
\end{array}
$$

Oxygen is considered, in itself, to be nonflammable. However, liquid oxygen has been known to react so vigorously with some fuel materials that an explosion results. Many substances that are essentially noncombustible, such as iron, will burn in oxygen.

It might be said that while gaseous oxygen promotes combustion in many materials that are not normally considered flammable, liquid oxygen, as a concentrated form of the gas, promotes combustion in those and many other materials, often at a greatly accelerated rate.

Liquid oxygen spills are especially hazardous because the liquid will penetrate even slightly porous materials, often remaining dormant for many minutes, just waiting for the slightest energy input to provide ignition resulting in violent flames or even explosion. Cases have been reported where just dropping a hose butt onto concrete has been enough to set off a violent explosion in such a situation. This phenomena of liquid oxygen remaining in porous materials has been observed with wood, asphalt, and powdered metal as well as concrete. Liquid oxygen is especially hazardous when clothing becomes wet from a splash or spill. The clothing must be removed and kept away from any possible ignition source for at least several hours.

Cutting off the liquid supply is the only remedy for fires involving liquid oxygen. Above all keep the cylinders cool with large amounts of water. Liquid oxygen can expand almost 900 times with a temperature rise of $200\,°F$. This expansion is enough to rupture the cylinders causing real trouble.

ACETYLENE, C₂H₂

Also called ethine and ethyne, acetylene is a colorless gas with an aroma like garlic and the following characteristics:

vapor density: 0.9
boiling point: −119 °F (−84 °C)
flash point: 0 °F (−18 °C) (cc)
autoignition point: 570 °F (299 °C)
ignition limits: upper 80%
lower 2.5% (by volume)

While acetylene is not prone to spontaneous ignition, it is nevertheless a great potential hazard, especially in an enveloping fire. The broad explosive limits make it particularly sensitive to flames that can lead to an explosion. The flame temperature of an acetylene-air mixture can be as high as 4220 °F (2327 °C).

In an enveloping fire, falling cylinders of this material have been known to break off their neck connections, ignite, and take off like rockets. They will travel several hundred yards and do considerable damage upon landing.

Acetylene is stored in *red* labeled cylinders.

Fires can be extinguished by stopping the flow of gas and applying a fine water spray, carbon dioxide, or dry chemicals.

CYCLOPROPANE, $CH_2 CH_2 CH_2$

Also called trimethylene, cyclopropane is a colorless gas, used in health care systems as an anesthesia. It has the following characteristics:

vapor density: 1.5
boiling point: −30 °F (−34 °C)
flash point: not available
autoignition point: 928 °F (498 °C)
ignition limits: upper 10.5%
lower 2.5% (by volume)

This is a very dangerous gas, the narcotic effects of which can be a serious hazard when it is out of control.

Cyclopropane has been known to react explosively when exposed in an enveloping fire, and will react violently with most oxidizing materials.

Cyclopropane is shipped in *red* labeled containers.

Fires can be extinguished by stopping the flow of gas and applying a fine water spray to the flames. Dry chemicals or carbon dioxide may also be used.

ETHANE, $C_2 H_6$

Also called dimethyl, bimethyl, methylmethane, and ethyl hydride, ethane is a colorless, odorless gas with the following characteristics:

vapor density: 1.04
boiling point: 129 °F (54 °C)
autoignition point: 960 °F (516 °C)
ignition limits: upper 12.5%
lower 3.0% (by volume)

This is a highly flammable gas that can be very dangerous in an enveloping fire. It will react vigorously with oxidizing materials to produce very high flame temperatures. It has been known to explode when collected in a confined area and ignition occurs.

Ethane is shipped in *red* labeled containers.

Fires can be extinguished by stopping the flow of gas and applying dry chemicals or carbon dioxide.

ETHYLENE, CH_2CH_2

Also known as ethene, etherin, and infrequently as elayl, ethylene is a colorless, pleasant-smelling gas that has the following characteristics:

vapor density: 0.98
boiling point: −155 °F (−104 °C)
autoignition point: 840 °F (449 °C)
ignition limits: upper 32%
lower 3.0% (by volume)

This is a very dangerous, very reactive, gas that burns or explodes with very little input energy. In confined areas it is a strong anesthetic. Ethylene is shipped in *red* labeled containers.

Fires must be extinguished only with great care. The flow of gas must be carefully stopped before any other remedy is attempted; then, a fine water spray may be applied. Dry chemicals or carbon dioxide may also be used after the flow of gas has stopped.

HYDROGEN, H_2

Hydrogen is a colorless, odorless gas that is highly explosive, with two isotopes that may be either natural or artificial. One is deuterium, ^2H, and the second tritium, ^3H. Hydrogen has the following characteristics:

vapor density: 0.07
boiling point: −422 °F (−252 °C)
autoignition point: 1085 °F (585 °C)
ignition limits: upper 74%
lower 4.0% (by volume)

Probably the most hazardous of the compressed gases, hydrogen's broad explosive limits make it a serious hazard potential. Even a weak spark of static electricity will ignite its mixtures within the ignition limits.

Hydrogen is shipped in *red* labeled containers.

Fires have been extinguished by cutting off the flow of gas and applying dry chemicals or carbon dioxide.

METHANE, CH_4

Also known as cooking gas, marsh gas and methyl hydride, methane is a color-less, odorless, tasteless gas that has the following characteristics:

> vapor density: 0.6
> boiling point: $-260\,°F\ (-162\,°C)$
> autoignition point: $1000\,°F\ (538\,°C)$
> ignition limits: upper 14%
> lower 5.0%   (by volume)

This very dangerous gas will burn vigorously in air and in an enveloping fire it may explode with great violence. In contact with chlorine gas, methane will explode when merely exposed to sunlight. Just mixing methane and fluorine gases together will produce an explosion even in total darkness.

Methane, like so many other compressed gases that are not anesthetic, is an asphyxiant by simple exclusion of oxygen from the body.

Methane is shipped in *red* labeled containers.

Fires involving methane can be extinguished with dry chemicals or carbon dioxide *after* the flow of gas has been stopped.

OTHER COMPRESSED GASES

There are several other compressed gases that find use in industrial and general service applications. These gases are not usually considered flammable although they may support combustion.

A few of the more common ones are:

Name	Formula	Vapor Density	Boiling Point
argon	A	1.4	$-300\,°F\ (-184\,°C)$
carbon dioxide	CO_2	1.5	sublimes at $-110\,°F\ (-79\,°C)$
helium	He	0.14	$-450\,°F\ (-268\,°C)$
neon	Ne	0.9	$-475\,°F\ (-282\,°C)$
nitrogen	N	0.95	$-320\,°F\ (-196\,°C)$
nitrous oxide	N_2O	1.5	$-130\,°F\ (-90\,°C)$

LIQUEFIED GASES

There are many gases, used for industrial work, whose critical temperature is high enough to allow the gas to be liquified at normal temperatures and handled in special cylinders. The gas is liquefied and put into cylinders where, at normal temperatures, the liquid boils off increasing the pressure of the vapor phase. When the vapor phase pressure is high enough, the boiling stops. When gas is drawn off for use the pressure is lowered and the boiling resumes until the cycle is completed.

Hazards develop, of course, when in an enveloping fire the temperature of the liquid inside the cylinder continues to rise until the vapor pressure approaches and surpasses the stress limitations of the container. When rupture does occur, not only is there danger of shrapnel, but the entire mass of liquid suddenly boils away releasing a tremendous quantity of gas.

There are those who would insist that only cryogenic gases can lead to a bleve. Anyone who has ever witnessed a liquefied gas tank rupture with the resultant fireball from the expanding gases would hardly argue the slight distinction. A boiling liquid is a boiling liquid no matter at what temperature the liquid boils at. There may be a difference in rate but it will be difficult to see it in the size of the fireball. An expanding gas fireball doesn't care whether it came from a cryogenic gas or not. In short, a bleve would seem to be a bleve as long as it comes from a boiling liquid.

Ammonia, NH_3

Ammonia is a colorless but very sharp smelling gas with the following characteristics:

vapor density:	0.6
boiling point:	$-28\,°F\,(-33\,°C)$
autoignition point:	$1205\,°F\,(652\,°C)$
ignition limits:	upper 25% (by volume) lower 16%

While ammonia is not usually considered flammable, it will burn if the gas concentration in air is high enough. It will also react explosively with compounds of silver and mercury. The greatest hazard from this gas, however, is not from flame or explosion, but from the toxic effects it has on the body. Ammonia gas is very soluble in water forming ammonium hydroxide, a caustic liquid. Ammonia attacks the eyes, mucous membranes, and skin, forming ammonia hydroxide with body water.

Ammonia gas must not be confused with household ammonia. That liquid is really a weak aqueous solution of ammonia hydroxide. It is sometimes called aqua ammonia.

Ammonia, therefore, may appear in three different forms: as ammonia gas, as anhydrous ammonia (liquefied ammonia gas), or as aqua ammonia.

Fires with ammonia gas are not uncommon, although it is considered nonflammable by the DOT. Such fires may be extinguished by first cutting off the flow of gas and then using a fine spray of water. Dry chemicals or carbon dioxide may also be used.

Butadiene, $CH_2CHCHCH_2$

Also known as erythrene, butadiene 1,3, and inhibited or uninhibited butadiene, butadiene is a colorless gas that has the following characteristics:

> vapor density: 1.87
> boiling point: 24 °F (−4 °C)
> autoignition point: 805 °F (429 °C)
> ignition limits: upper 11.5%
> lower 2.0% (by volume)

Butadiene is an extremely dangerous compound that flames with relative ease. It may also form very unstable, explosive peroxides by mere exposure to the atmosphere. It is shipped in *red* labeled containers.

Because of its low boiling point of 24 °F (−4 °C), it evaporates very rapidly at normal temperatures, absorbing a considerable amount of heat. If, therefore, butadiene is spilled on or contacts the skin in any way, its rapid evaporation can cause frostbite.

Fires can be extinguished by cutting off the gas flow and using a fine water spray, carbon dioxide, or dry chemicals.

Butane, $CH_3CH_2CH_2CH_3$

Also known as n-butane, methylethylmethane, and butyl hydride, butane is a colorless, practically odorless, gas with the following characteristics:

> vapor density: 2.1
> boiling point: 31 °F (−1 °C)
> autoignition point: 761 °F (405 °C)
> ignition limits: upper 8.5%
> lower 1.9% (by volume)

Butane is a highly dangerous gas that is very flammable. It will explode in an enveloping fire with very little input energy.

Butane is one of the liquefied petroleum gases that have become so popular in the last few years for heating, cooking, and industrial uses. It may be packaged in any one of several different sized cylinders, from small one-pound tanks to those holding many thousands of gallons for industrial uses.

With a vapor density of 2.1 and high cohesive forces between molecules, butane tends to form a cloud that stays close to the ground until it finds an ignition source.

Butane, for consumer use, may be mixed with sulfur or mercaptans to provide an odor that should give warning of a leak. In industrial applications, where pure butane is needed, no mercaptan can be added and so a leak is not as easily detected.

A small fire with butane may be fought only by stopping gas flow, then applying dry chemicals to the base of the flames.

THE HALOGENS

Some of the most common and chemically active elements belong to the same family, the halogens. The members of the halogen family, in ascending order of atomic weight, are fluorine, chlorine, bromine, iodine, and astatine, a very rare element. Of these, only fluorine and chlorine are gases at normal temperatures. However, many organic compounds of bromine can be found as hazardous gases.

In general, the halogen group is an extremely reactive and toxic family of elements. Their reactivity is attested to by the great number and wide range of halogen compounds found in everyday life. They vary from insecticides to refrigerants, from extinguishing agents to highly dangerous explosives.

All halogen gases form corrosive acids with body liquids so protective clothing must cover every inch of the body during exposures to these materials. The halogen gases are heavier than air but will travel great distances on the wind, wreaking havoc all along the path.

Fluorine, F_2

Fluorine, a pale yellow gas, should not be confused with fluorene, which is a white, relatively inactive organic solid. Fluorine is a member of the halogen family and is very reactive, with an unmistakably pungent odor and the following characteristics:

vapor density: 1.7
boiling point: $-300\,°F\ (-184\,°C)$

While fluorine does not have a flammability point as such, it is a very reactive gas and is extremely hazardous. It will react with just about all known inorganic and organic materials. Because it is such an excellent oxidizing agent, it is a real fire hazard. It is shipped in *red* labeled containers.

In an enveloping fire, when heated, it becomes even more reactive, especially with water in any form. Steam and heated fluorine gas react to produce highly corrosive liquids of hydrofluoric and hydrofluorous acid. These attack the mucous membranes of the body, as does fluorine itself. Burns resulting from contact with fluorine or the acids, hydrofluoric and hydrofluorous, are very serious, take a very long time to heal, and, more often than not, leave permanent scars.

Hydrofluoric Acid, HF

Also known as hydrogen fluoride and fluorhydric acid, hydrofluoric acid is a clear, colorless gas that may also exist as a highly corrosive liquid. It has the following characteristics:

vapor density: not available
boiling point: 67 °F (19 °C)

Hydrofluoric acid is considered a nonflammable material, but it is one of the most hazardous gases in use today. It is a very corrosive substance, and because it reacts so readily with body fluids, concentrations as low as 50 ppm, even for short periods, should be considered dangerous.

Burns from this material must be treated as soon as possible not by just washing with water, but by *long soaking in neutralizing solutions,* such as Epsom salts. These burns should be seen by a physician as soon as possible.

In an enveloping fire, hydrogen fluoride decomposes to produce other very toxic fluoride compounds.

It is shipped in *white* labeled containers.

Chlorine, Cl₂

This is normally found as a greenish yellow gas and is a very common, much used element. It has the following characteristics:

vapor density: 2.5
boiling point: 30 °F (−1 °C)

Because of that 30 °F (−1 °C) boiling point, chlorine may also be found as a liquid and, under proper conditions, as a crystalline solid.

Chlorine is a very reactive material that may in certain cases react so violently that it will seem to explode.

The mere contact of this gas, even at normal temperatures, with some hydrocarbons, such as turpentine or illuminating gas, may produce violent explosions.

When heated in an enveloping fire, it will react with water and especially steam, to produce the highly corrosive liquids, hydrochlorous and hydrochloric acids, as well as the very corrosive vapors of these chlorides.

It is not considered a flammable gas and is therefore shipped in containers bearing *green* labels.

Organic Halogens

The organic halogens are in general narcotic substances but of varying strengths. They are dangerous materials when exposed to heat or flames and, in an enveloping fire, decompose to very toxic chloride, fluoride, or bromide fumes.

Fires have been extinguished by stopping the flow of gas and then applying dry chemicals or carbon dioxide.

A few of the more common organic halogen gases in use today are:

Name	Other Name(s)	Formula	Vapor Density	Flash Point	Autoignition Point	Ignition Limits Upper	Ignition Limits Lower
methyl chloride	chloromethane	CH_3Cl	1.8	32 °F (oc) (0 °C)	1170 °F (632 °C)	18%	9%
ethyl chloride	chloroethane hydrochloric ether muriatic ether	CH_3CH_2Cl	2.2	−59 °F (cc) (−51 °C)	965 °F (518 °C)	15%	4%
vinyl chloride	chloroethylene chloroethene	CH_2CHCl	2.15	−108 °F (oc) (−78 °C)	882 °F (472 °C)	22%	4%
methyl bromide	bromomethane	CH_3Br	3.3		998 °F (537 °C)	15%	12%
ethyl bromide	bromethane hydrobromic ether bromic ether	CH_3CH_2Br	3.76		951 °F (511 °C)	11.2%	7%

REVIEW QUESTIONS

1. Define "gas." How does a gas differ from a material in the other states of matter?
2. Give the three major characteristics of the gaseous state.
3. Define "critical temperature" and "critical pressure." Which is more important to a firefighter?
3. What is a bleve? How does it occur?
4. Discuss cryogenics and cryogenic materials. What are the hazards of cryogenics?
5. What is the difference between inert and active gases?
6. Discuss the general hazards that can be expected with liquefied gases.
7. What are the halogens? Why are they considered hazardous?
8. Discuss the difference between compressed and liquefied gases.
9. Why is acetylene such an active compound?
10. Describe your choice as the most hazardous of gases and give reasons for your decision.

Hazardous
Material Warnings/ 9

Separating the good from the bad can be difficult in any work, but without external aids it can be almost impossible in a fire or potentially hazardous situation. For this reason, several systems or methods of classification have been set up to be used by those responsible for a given material to warn those who may have to deal with it in a hazardous situation.

These warnings may take several forms but the most popular are:

- Signs both large and small carrying warnings of all sorts. They can exaggerate danger or underplay it. Too often such signs may be used to discourage trespassers, but fire safety personnel cannot afford to treat them lightly. Take them at their face value.
- System placards and signs. Two well planned and informative systems are recognized at this time, The Department of Transportation (DOT) system for moving vehicles and The National Fire Protection Association 704M system for fixed installations.

LEVELS OF HAZARD

Both the NFPA and DOT systems are developed around basic definitions of the hazards of a material. The four levels of hazard are recognized:

1. extremely hazardous
2. very hazardous
3. mildly hazardous
4. hazardous

EXTREMELY HAZARDOUS

Compounds or combinations of compounds that are basically unstable and extremely flammable, which can cause death or serious injury even after very short exposure, are labeled **extremely hazardous**. These materials may be radioactive, toxic, flammable, corrosive, or explosive, and belong in this class as long as the adjective "highly" (or extremely) must be used before each term to properly describe its level of hazard.

VERY HAZARDOUS

In general, materials with high chemical reactivity but more stability than those of the extremely hazardous class are labeled **very hazardous**. Such materials require more energy input to upset them and begin reactions. They may not cause death on short exposure but have been known to do so under longer exposure, and they usually are more disabling than lethal. These materials may also be radioactive, as long as the term "moderately" should be used to properly describe their level of hazard.

MILDLY HAZARDOUS

Mildly hazardous materials require external energy input to initiate a hazardous condition. Low level radioactive materials requiring long term exposure, as long as there are detrimental results, belong in this class, as do toxic materials that produce temporary disability.

HAZARDOUS

All other materials that are capable of causing temporary discomfort or disability or that are combustible with considerable external energy input are classed as **hazardous**.

THE DOT SYSTEM

The Department of Transportation Hazardous Warning System, required for transporting vehicles, consists of large diamond-shaped placards usually just over 14 inches on each side with various colored backgrounds, stripes, lettering, and silhouettes to indicate the type and level of hazard present.

The system is broken down into two basic classifications: those shipments of any quantity that require certain placards; and those shipments of given quantities or special situations that require certain placards.

ANY QUANTITY

Placards are used on motor vehicles, freight containers, and rail cars containing any quantity of certain materials, listed below. These placards are illustrated in Figure 9.1.

Hazardous Material Classed or Described as	*Placard*
Class A explosives	EXPLOSIVES A
Class B explosives	EXPLOSIVES B
Poison A	POISON GAS
Flammable solid (dangerous when wet)	FLAMMABLE SOLID W
Radioactive material	RADIOACTIVE
Radioactive material uranium hexafluoride, fissile (containing more than 0.7% U^{235}	RADIOACTIVE and CORROSIVE
Radioactive material uranium hexafluoride, low specific act. (containing 0.7% or less U^{235}	RADIOACTIVE and CORROSIVE

background: orange
printing: black
silhouette: black

background: orange
printing: black
silhouette: black

Figure 9.1.

background: white
printing: black
silhouette: black

stripes: red and white
top triangle: blue
printing: black
W : white

background: white
top triangle: yellow
printing: black
silhouette: black

Figure 9.1. (continued)

OTHER REQUIREMENTS

For motor vehicles, rail cars, and freight containers containing 1000 pounds (454 kilograms) or more gross weight of hazardous materials placards are required, as illustrated in Figure 9.2.

Hazardous Material Classed or Described as	*Placard*
Class C explosive	FLAMMABLE
Nonflammable gas	NONFLAMMABLE GAS
Nonflammable gas (chlorine)	CHLORINE
Nonflammable gas (fluorine)	POISON
Nonflammable gas (oxygen, pressurized liquid)	OXYGEN
Flammable gas	FLAMMABLE GAS
Combustible liquid	COMBUSTIBLE
Flammable liquid	FLAMMABLE
Flammable solid	FLAMMABLE SOLID
Oxidizer	OXIDIZER
Organic peroxide	ORGANIC PEROXIDE
Poison B	POISON
Corrosive material	CORROSIVE
Irritating material	DANGEROUS

background: red
printing: white
silhouette: white

background: green
printing: white
silhouette: white

Figure 9.2.

background: yellow
printing: black
silhouette: black

background: yellow
printing: black
silhouette: black

background: white
printing: black
silhouette: black

background: red
printing: white
silhouette: white

background: black
printing: white
silhouette: black

backgrounds: red/white/red
printing: black

Figure 9.2. (continued)

background: red
printing: white
silhouette: white

stripes: red and white
printing: black
silhouette: black

Figure 9.2. (continued)

MIXED SHIPMENTS

When a freight container, rail car, or motor vehicle contains *two or more* classes of hazardous materials, requiring two or more different placards as noted above, the DANGEROUS placard may be used in place of the separate placards specified for each class. *However,* when 5000 pounds (2270 kilograms) or more of *one* class of hazardous material is loaded at one loading facility, the placard for that class must be applied.

OTHER GASES

The OXYGEN placard (Figure 9.3) is used for pressurized liquid oxygen. However, this placard may also be used to identify liquefied oxygen contained in a manner that does not meet the definition of a compressed gas as given in this chapter.

CHLORINE

The CHLORINE placard, also shown in Figure 9.3, is used only for packaging having a rated capacity of more than 110 gallons (416 litres). The NON-FLAMMABLE GAS placard is used for packagings having a rated capacity of 110 gallons (416 litres) or less.

background: yellow
printing: black
silhouette: black

background: white
printing: black
silhouette: black

Figure 9.3.

HIGHWAY SHIPMENTS

The GASOLINE placard may be used in place of the FLAMMABLE placard when gasoline is being transported. In like manner, the FUEL OIL placard may be used in place of the COMBUSTIBLE placard when fuel oil that is not in the flammable liquid category is being transported. These placards are shown in Figure 9.4.

background: red
printing: white
silhouette: white

background: red
printing: white
silhouette: white

Figure 9.4.

EMPTY RAIL SHIPMENTS

Because flammable vapors are so hazardous, and because empty tank cars may contain dangerous ratios of air and flammable vapors, an EMPTY placard (Figure 9.5) is required that corresponds to the placard that was required for that material the tank car last contained unless that last material was a combustible liquid.

The EMPTY placard is required for the following hazardous materials: nonflammable gas, flammable gas, oxygen, chlorine, poison gas, flammable, flammable solid, flammable solid W, oxidizer, organic peroxide, poison, and corrosive.

backgrounds:
top: black
bottom: red
printing: white

Figure 9.5.

OTHER DOT WARNINGS

The Department of Transportation has several labels for use on hazardous material packages. These labels are intended to warn emergency personnel of the material or materials within the package. While the labels are, for the most part, similar to the placards used for transportation vehicles, they are smaller, about 4 inches by 4 inches, compared to the 14 inches on a side for the placards.

These labels are supposed to appear on each package and, should multiple packages be combined into large single cases, the outside case should carry all the warnings of each of the individual packages.

In addition to the symbols found on placards, the labels may carry:

• ETIOLOGIC AGENT: This label (see Figure 9.6) warns of biomedical material. The original meaning of the word "etiology" is the science of causes or reasons why things occur. The medical use of this word has come to mean that branch of medicine that deals with the causes of diseases. The contents of these packages can be deadly. The etiologic symbol looks like the surrealistic portrayal of the pincers of some particularly nasty disease germ ready to bite.

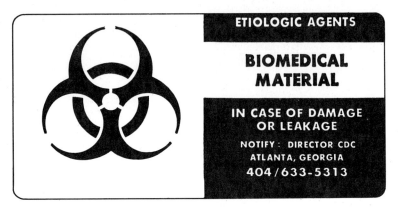

red and white

Figure 9.6.

· SPONTANEOUSLY COMBUSTIBLE: For extremely unstable material, this label (Figure 9.7) must be used with other applicable label(s) in order to be more specific. This label has a white triangle on top with a black flame silhouette in it and a red triangle on the bottom with the words "Spontaneously Combustible" printed in black.
· DANGEROUS WHEN WET: This label (also seen in Figure 9.7) is for water reactive material that must be kept dry. It should be used with other labels in order to be more specific. This label has a blue background with a black flame silhouette in the upper section. The words "Dangerous When Wet" are printed in black just below the flame silhouette.

The preceding DOT classifications contained many categories that require definition. They are listed alphabetically below.

Class A explosive: any material that detonates or is a maximum hazard

Class B explosive: flammable hazards that are dangerous because of rapid combustion

Class C explosive: minimum hazard, but may contain restricted amounts of class A or class B explosives

Combustible liquid: any liquid having a flash point at or above 100 °F (38 °C) but below 200 °F (93 °C)

backgrounds:
top: white
bottom: red
silhouette: black

background: blue
printing: black
silhouette: black

Figure 9.7.

Corrosive material: a solid or liquid material that causes visible deterioration or irreversible alteration in human tissue at the point of contact, or a material that causes severe deterioration in a metal with which it comes in contact

Flammable gas: a compressed gas that can be ignited

Flammable solid: any solid, other than those which are in the explosive category, that under conditions normally found in transportation is liable to cause fires by means of friction or retained heat from manufacturing or processing, or that can be ignited readily and when ignited burns so vigorously and persistently as to create a serious transportation hazard

Flammable solid (dangerous when wet): a solid material that will spontaneously react with water to become flammable or produce a gas that is flammable or toxic or both

Flammable liquid: any liquid having a flash point below 100 °F (38 °C)

Irritating material: a liquid or solid material that when exposed to air or when subject to flame develops vapors that produce undesirable effects on living tissue

Nonflammable gas: any material or mixture in a container developing absolute pressure greater than 40 psi (280 kPa) at 70 °F (21 °C) or, regardless of the pressure, at 70 °F (21 °C), having an absolute pressure greater than 104 psi (728 kPa) at 130 °F (54 °C)

Organic peroxide: any compound that contains the peroxide linkage (—O—O—) and that has one or more hydrocarbon groups replacing the normal hydrogen atoms

Oxidizer: a substance that readily produces oxygen to stimulate the combustion of organic matter

Poison A: extremely dangerous poisonous gases or liquids of such nature that a small amount of the gas or vapor from the liquid, when mixed with air, is dangerous to life

Poison B: poisonous liquids or solids (including pastes or semiliquids) which are chiefly dangerous by external contact with the body or by being taken internally as in contaminated food or seeds. The vapors of some of this class of poison are also offensive and dangerous but to a lesser extent than class A poisons.

THE NFPA 704M SYSTEM

While the DOT system fills a definite need to warn fire safety personnel of hazards in moving containers, it is rarely used on fixed installations. The National Fire Protection Association No. 704M System of Hazard Identification fills that need for warnings on fixed installations very well. This system codes the levels of hazards by number and therefore is useless to untrained personnel. The coding, however, is not difficult and should be a must for all interested in fire safety.

The 704 system makes use of four diamonds arranged to form one large diamond, and this, together with the multiple coding, allows the warnings considerably more flexibility. While the 704 system provides more information in one sense, in another, it does not specify the exact type of hazard as does the DOT. It would be much better for all concerned if both systems were demanded for all hazardous installations and moving vehicles transporting hazardous materials.

The large 704 diamond is divided into four blocks or smaller diamonds with each one reserved for a specific bit of information important to fire safety personnel (Figure 9.8).

If we consider the 704 diamond as if it were a baseball field, the square at first base, which may have a yellow background, indicates the chemical stability of the material. The stability is divided into five levels with the number 4 indicating highest reactivity and 0 the lowest, as follows:

4—This material will almost certainly detonate with very little provocation. Clear the area. Don't fight this fire.

3—The material may detonate when heated and confined. Evacuate the area and do not fight this fire with hand-held lines.

2—These materials may undergo very rapid chemical reaction at higher than normal temperatures and pressures. The reaction may be so violent that it seems like a detonation. Keep a good distance away if attempting to cool with hand lines. Use extreme caution.

1—The material is usually stable at normal temperature and pressure, but may be triggered to instability by other materials or when raised to

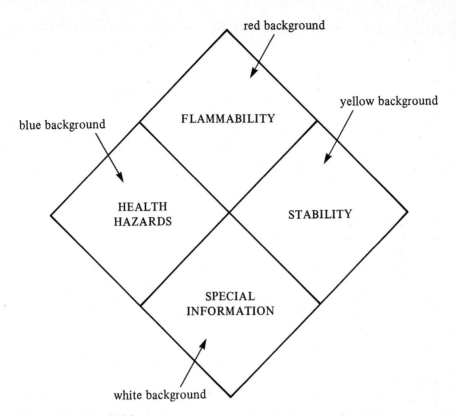

Figure 9.8. The 704 System.

elevated temperature and pressure. Take normal precautions to cool or extinguish.

0—The material is stable and should not produce chemical activity hazard for fire safety personnel.

The square at second base, which may have a red background, indicates the tendency of the material to burn. There are five levels to this block, again with the number 4 indicating the highest tendency and 0 the lowest, as follows:

4—This material is highly flammable; the number 4 is usually used for very flammable gases and highly volatile liquids. Use extreme caution.

3—This material may ignite under normal temperature conditions with very little external energy input.

2—This material must receive a moderate amount of external heat before it will flame.

1—This material must receive considerable external heat input before it will ignite.

0—This material will not burn.

At third base, which may have a blue background, the number indicates the health hazard of the material, and it too is coded into five levels. Again, 4 is the greatest health hazard and 0 the least, as follows:

4—This material is far too dangerous for even the slightest exposure. Ordinary protective clothing is *not* adequate. Get special help.

3—This material is extremely hazardous, but full protective gear with special precautions to prevent any contact whatsoever should be acceptable. Use extreme caution.

2—This material is very hazardous but use of self-contained breathing apparatus will allow entry into the area.

1—This material is a hazard only if body is directly exposed and no treatment follows. Minor irritation will result.

0—This material is no hazard but may be combustible.

The home plate block, with a white background, is used to relay special information. Originally only two symbols were used:

- W indicates that the material may react violently with water. It does *not* mean "do not use water". It means that if you must use water be very careful—get more information before proceeding. (See CHEMTREC, under Special Assistance in this chapter.)
- The radioactive propeller indicating that the material is radiologic in nature. See Chapter 13.

As the system evolves, however, more symbols are being added. Whether they are official or unofficial is not important—these new symbols communicate valuable information.

Three of the newer symbols are:

- The BIOMEDICAL warning is a modification of the DOT etiological placard that may be found in the fourth square (Figure 9.9). In hazardous situations handle these containers only with expert advice.
- OXY in the fourth square warns of oxidizer materials. These can be serious fire or disaster hazards.

red on white

Figure 9.9. The Biomedical Warning.

- COR in the fourth square has been used to indicate corrosive materials. The slightest indication of a leak or spill warns of serious hazards.

While we have used baseball terminology to simplify the presentation, the numbers are more often than not read from left to right putting the health block first, the flammability block second, the stability block last. For example, sodium hydroxide would be read 3-0-1.

There is considerable flexibility used in actual practice with the 704 diamond. It does not always appear as a four-square diamond, nor do the individual blocks always have colored backgrounds. The level-of-hazard numbers may appear without the home plate block as in Figure 9.10.

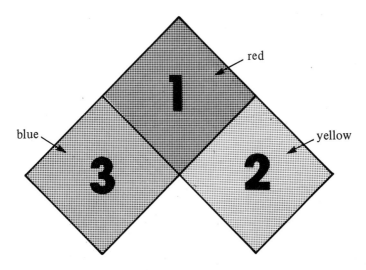

Figure 9.10. Modified 704 System.

The numbers may or may not have a background but usually, when there is no diamond background, the numbers themselves may be yellow, red, and blue in the proper positions. Sometimes the numbers are presented as a partial diamond as in Figure 9.11.

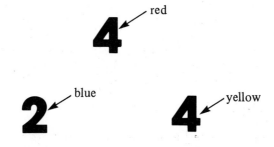

Figure 9.11. Partial Diamond.

While the NFPA 704 system can warn of different levels of hazards in several different areas, it is not specific. It does not, for example, warn that a material is an oxidizer as such. If there is any doubt, seek special assistance.

Other methods of presenting the 704 system are shown in Figure 9.12 and Figure 9.13.

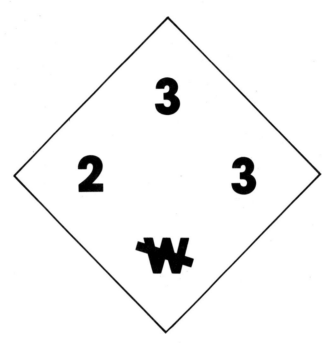

Figure 9.12.

SPECIAL ASSISTANCE

CHEMTREC

The Manufacturing Chemists Association, at 1825 Connecticut Avenue, N.W., Washington, D.C., provides assistance in any transportation emergency involving chemicals within the continental United States. This service is called CHEM-TREC, which stands for Chemical Transportation Emergency Center.

This public service provides immediate advice for those at the scene of emergencies, then promptly contacts the shipper of the chemicals involved for more detailed assistance and follow-up. CHEMTREC is *not* a general information source, but is designed to deal with chemical emergency situations.

CHEMTREC operates 24 hours a day, 7 days a week, to receive direct-dial, toll-free calls from any point in the United States. The telephone numbers are:

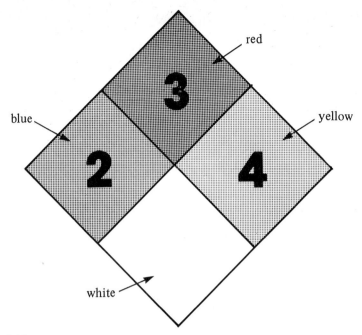

Figure 9.13.

within the District of Columbia, 202-483-7616; and from all other locations, 800-424-9300.

IRAP

For radiologic emergencies, monitoring help may be obtained through the *Inter-agency Radiological Assistance Plan*, which has offices throughout the country. Calling the nearest office will provide expert information both during and after radiologic emergencies. The telephone numbers are:

Region	Regional Office	Telephone Number
1	New York, NY	516-345-2000
2	Oak Ridge, TN	615-483-8611
3	Savannah River, GA	803-824-6331
4	Albuquerque, NM	505-264-4667
5	Chicago, IL	312-739-7711
6	Idaho Falls, ID	208-526-0111
7	San Francisco, CA	415-841-5121
8	Richland, WA	509-942-7381

NOTE: Canal Zone should call Savannah River
Alaska should call Richland
Hawaii should call San Francisco
Puerto Rico and the Virgin Islands should call Oak Ridge

REVIEW QUESTIONS

1. Name the two principal warning systems and detail the difference between the two.
2. What is the basic difference between extremely hazardous and very hazardous?
3. What is the basic difference between mildly hazardous and hazardous?
4. Why are class C explosives labeled flammable in the DOT system, while class A and class B explosives must be labeled as class A or class B?
5. When is the DANGEROUS placard used? What does it mean?
6. What should the EMPTY placard mean to you as part of a fire fighting team at the scene of a tank car fire?
7. Summarize the NFPA 704M system. What are some variations on it?
8. In the NFPA system, what does a number 4 indicate in any box?
9. What does the slashed W in the bottom block indicate in the NFPA system?
10. What is CHEMTREC? What is its function?
11. What is IRAP? How can it help you?

Transportation Hazards/ 10

The word "transportation" indicates the movement of a mass. Movement of a mass, however, requires energy. When this energy is under control, it works well for us; when it is out of control, it can contribute to hazardous situations whether they are on land, on water, or in the air.

The energy involved in the movement of mass is usually transformed from potential, chemical energy to kinetic, mechanical energy in a device called an engine. The energy is derived from a fuel.

FUELS

There are solid fuels (such as coal), gaseous fuels (such as propane) and liquid fuels (such as gasoline). Because of their characteristic efficiency and the relative ease of extracting their energy, liquid fuels are, by far, the most popular. Because they are so plentiful, contain considerable energy, and release it with relative ease, they are the greatest hazards.

LIQUIDS

The more popular liquid fuels are gasoline, kerosine, and diesel oil.

Chemically speaking, the fuel with the least energy is gasoline. However, because of the quantities available and the hazardous characteristics of gasoline, it has a very great potential for disaster.

The gasoline we use every day is essentially an isomeric mixture of heptanes (C_7H_{16}). The energy release which makes it, as well as any fuel, useful is called the **heat of combustion**. It is usually expressed in terms of kilogram calories per mole (Kg-cal/mole). However, it can be expressed in Kg-cal/gallon in order to make it more meaningful to us. A comparison of heats of combustion for a few of the common gaseous and liquid fuels is shown below.

Compound	Number of Carbons	Kg-cal/gal	BTU/gal
methane	1	1450	5753.6
propane	3	1810	7182.1
hexane	6	1970	7816.9
heptane	7	2040	8094.7
octane	8	2090	8293.1

It doesn't take much study of the above data to see that increase in chain length is correlated with an increase in energy. This is not directly due to the length of the chain, but to the hydrogen content of the molecule. All of the hydrocarbons listed in the above table are saturated, that is, there are no carbon double bonds in the molecule. Each carbon has as many hydrogen atoms attached to it as the structure will allow. The heat of combustion of hydrogen is 15.4 kilocalories per pound (33.9 kilocalories per kilogram). Carbon's heat of combustion is only 3.67 kilocalories per pound (8.1 kilocalories per kilogram).

While there are only 4 hydrogen atoms per molecule with methane, there are 18 hydrogens in octane.

Gasoline comes from the distillation fraction of petroleum that contains the C_7 to C_{10} groups, kerosine from the C_{11} to C_{16} fraction, and diesel oil from the C_{15} to C_{20}. It is not difficult to see, therefore, that the diesel oils have the higher heat of combustion because they have more hydrogen bonded to the longer carbon chains.

In actual practice, however, gasoline is modified in order to increase its combustion energy. This is accomplished by adding other liquids such as tetra-ethylead, $(C_2H_5)_4Pb$, in order to control burning, and isooctane (C_8H_{18}). It is from the latter that the octane rating of gasoline is derived.

The octane rating of gasoline is developed by comparing the knocking or pinging qualities of a given gasoline formula in a test engine with that of another formula of gasoline containing varying percentages of heptane and isooctane. For example, a 90 octane gasoline is a mixture that has the same pinging properties in a standard test engine as does a mixture of 10 percent heptane and 90 percent isooctane.

Kerosine

While kerosine will produce energy in a diesel engine, it is not normally used in this way because of its cost and relative efficiency. Kerosine may be used in a burner at atmospheric pressure to heat water in a steam engine. However, even in this application, the heavier fuel oils are more efficient and far less expensive.

Kerosine is used extensively in mixtures of jet aircraft turbine fuel. Because these mixtures are unique blends they will be covered at length in the aircraft fuel section of this chapter.

Fuel Oils

"Fuel oil" is a rather general term that is used for a multitude of products. Chemically, they are long chain carbon compounds containing 20 to 30 carbons in the chains with hydrogen, oxygen, sulfur, and dissolved minerals.

In actual use, there are a great many types of fuel oil with varying characteristics. Basically even fuel oils should be divided into two types, residual and distillate. Just as their names imply, the distillate type is gathered from the fractional distillation of crude oil while the residual is the remaining liquid product of the refining process. Large commercial installations, whether on land or in marine use, rarely employ the distillate types in burners for their steam boilers. The distillate types are better known as fuel oil number 1 or fuel oil number 2. These installations usually use the residual types, which are also known as fuel oil number 4, number 5, number 6, or bunker C. The rating number 3 is no longer used.

The residual oils are heavy, viscous liquids, that with the exception of number 4, usually require some degree of preheating in order to thin them out enough to allow them to be moved out of their holding tanks and into the burners. While grade number 4 will flow at temperatures as low as 20 °F (−7 °C), grade number 5 is practically a gel at 40 °F (4 °C) and below. Number 6 and bunker C require heating not only to improve their flow properties but for ignition as well.

In spite of these flow characteristics, these oils are combustible when sufficient input energy is available. Their flash points, however, are all above 130 °F (54 °C), so that the hazards caused by their volatility are minimal.

The term diesel oil is really just a specific classification for fuel oil number 2. Because the diesel engine has the ability to extract the energy from the less expensive distillate fractions, the highly refined gasoline product is not needed. Diesel oil is less expensive than gasoline or kerosine because it undergoes much less refining. The diesel engine has its disadvantages however: the fuel must be injected at high pressure in order to realize maximum efficiency. The pressurizing air for the fuel can therefore become a hazard in case of an accident. Laboratory experiments have shown that air under pressure as low as 15 psi (105 kPa), when injected through the proper nozzle and into a cavity in dry wood, can result in ignition of the wood.

Because of the differences in the operating principles of the diesel as opposed to the spark-ignition engine, low octane fuels operate much more efficiently in the diesel engine than do the high octane fuels. For this reason the ignition property of diesel fuel is *not* rated by octane number. Diesel fuels are rated by **cetane** number. Cetane is another name for n-hexadecane ($C_{16}H_{34}$).

In the determination of the cetane number, diesel fuel is compared to a mixture of cetane and a-methylnaphthalene. If a given diesel fuel requires the same input energy for ignition in a test engine as does a 47 percent cetane, 53 percent a-methylnaphthalene mixture, the fuel is given a cetane rating of 47.

SOLIDS

At the present time the number of solid fuels used in transportation engines is quite limited. Coal seems to have the field to itself. With the coming of the space age and the need of rocket fuels or the even more impending energy crisis, this may change, but for the present we shall limit ourselves to coal.

Coal

Even with the energy crisis there is an excellent chance that the large land and marine boiler systems may be forced to return to the old fashioned fuel, coal.

Coal is actually the result of long-term (millions of years) chemical reactions involving cellulose $(C_6H_{10}O_5)_x$. An oxidation-reduction reaction, with a limited supply of air but with an abundance of water, took place under the influence of pressure and above normal temperatures. The various stages of this reaction are as follows:

$$\text{peat} \longrightarrow \text{lignite} \longrightarrow \text{soft coal} \longrightarrow \text{hard coal}$$

Each of the above steps is characterized by an increasing carbon content. Soft coal contains about 80 percent carbon, 6 percent hydrogen, 5 percent oxygen, 2 percent nitrogen, 2 percent sulfur and about 5 percent mineral salts. Lignite, sometimes called brown coal, is the solidified form of lignin and cellulose. Lignin is the sticky, organic substance by which wood fibers are bound together.

Contrary to popular thinking, even hard coal is not pure carbon. In fact, carbon is more often than not combined with other chemical elements in the form of large polymers. Anthracite, a very hard coal, comes the nearest to being all carbon. But even here some small amount of combining has taken place.

Coal, as used for steam installations, has five main classifications:

1. **Anthracite** is a very high carbon, very low sulfur-bearing solid that burns clean and has a very low dust hazard during handling.
2. **Semianthracite** is a high carbon solid with more hydrogen and oxygen than anthracite. Semianthracite is not as hard as anthracite with the result that the dust hazard is slightly greater.
3. **Bituminous,** or soft, coal is really a catchall classification containing many different chemical combinations. It may have a high sulfur content, may burn with a great deal of smoke, and will often produce considerable solid, black residue (soot) if not carefully controlled during the combustion process. It has a very high dust hazard and has been known to undergo spontaneous ignition under some storage conditions. The dust is very sensitive to static electricity and spark ignition.
4. **Semibituminous** is a very powdery, soft coal which develops considerable dust. This dust is sensitive to all forms of ignition. Under some storage conditions, in bulk form, it has been known to ignite spontaneously.

5. **Subbituminous** coal is just one step above lignite, which has a woody structure, and therefore does not have a high dust problem. It will, nevertheless, ignite spontaneously under certain conditions.

GASES

Gases are not used very extensively for the transportation of large masses of cargo, although they may be used in special applications in industry.

With the realization that the hydrocarbon fuels derived from fossilized vegetation are in limited supply, more and more attention is being given to the development of engines using chemically produced gases such as methane and propane. Internal combustion engines with their spark ignition systems can be adapted to these gaseous fuels without extensive redesign. Propane and methane are treated in detail in the chapter on hazardous gases.

As with any gaseous fuel, methane and propane have broad explosive limits and will cause considerable damage if a combustion reaction goes out of control.

Serious consideration is also being given to hydrogen as a fuel. Its high heat of combustion, cleanliness of burning, and excellent efficiency, along with its ease of generation by relatively economical chemical reactions, make it an ideal candidate. However, the fact that it is a gas with such a high heat of combustion makes it an extremely hazardous material to handle. There would be a much greater chance of an explosion with hydrogen, in an accident, than with gasoline.

ENGINES

Engines, in themselves, are merely a mechanical system of moving parts. Normally it is the fuel coupled with the engine and the way the fuel is used in the engine that produces a potential hazard or hazards.

Every engine, no matter of what type, has a certain amount of inefficiency. This inefficiency appears to us as heat. While the heat can be used to make small improvements in the overall engine performance, there is still a considerable amount of heat stored in the metal parts and associated engine components. In case of an accident, where volatile fuel or low flash point cargo can come in contact with the engine, this heat energy can be an excellent ignition source.

THE DIESEL ENGINE

Developed by Rudolf Diesel in 1893, this engine makes use of the phenomenon of compression ignition. That is, during one of its upward strokes, a piston compresses the air that has been entrapped by proper timing of the valves. Near the top of this stroke, fuel, in the form of a spray, is injected into the cylinder.

As the compression of the fuel-air mixture is continued by the still upward movement of the piston, ignition occurs. The combustion reaction takes place because of the high temperatures (500 ° to 700 °F, 260 ° to 371 °C) developed by the compression of the mixture. The release of the chemical energy drives the piston downward, performing the work as required. In this manner the diesel extracts the chemical energy stored in the fuel and converts it into mechanical energy. The temperature in the cylinder can go as high as 4700 °F (2593 °C) with pressures as high as 1500 psi (10.5 MPa) after ignition and during the combustion process.

The very high conversion efficiency of diesel engines can be improved by injecting the air into the cylinders under greater pressure. This increases the oxidizer/fuel ratio and improves the output energy, and is commonly called "supercharging."

THE STEAM ENGINE

While the diesel engine converts the chemical energy of a fuel directly into mechanical energy, a steam engine or a steam turbine extracts chemical energy indirectly by means of a boiler. In the boiler, fuel is burned and chemical energy released in the form of heat. This heat is used to change the physical state of water from a low energy state (liquid) to a higher energy state (gas).

There are two types of energy converters using steam, the steam engine and a steam turbine.

The steam engine uses the energy developed by the steam to drive pistons back and forth within a cylinder. There may be more than one cylinder to an engine. The reciprocating motion is converted to rotary motion and this used to propel the transporter.

The steam turbine develops rotary motion directly by having multiple sets of wheels containing spine-like blades. These wheels are mounted on a single shaft. The steam enters at one end, and as it rushes toward the other end of the turbine, it pushes against the blades, turning the shaft and driving the transporter.

Whether the steam engine is of the reciprocating type or the turbine type, both depend on the boiler to provide the steam pressure that is used to perform the work of driving the transporter. Pressure is developed from conversion of the chemical energy of the fuel transforming the liquid into a gas, which, for water, occupies 1670 more volume than the liquid. In systems where superheated steam is used this number is much greater.

In a superheated system, more heat is added to the steam so that it expands even more than normal. Modern systems work with steam pressures well over 1000 psi (7 MPa) and at temperatures above 1382 °F (750 °C).

Normal, or nonsuperheated, steam is sometimes referred to as **wet steam** because it contains droplets of liquid water. Properly, superheated steam is referred to as **dry steam** because it is 100 percent gas and therefore contains more energy.

In either case, high pressure, together with high temperature, constitute a direct hazard for fire personnel. A sudden leak, as the result of an accident, can cause severe injury. High pressure and/or high temperature steam lines must be treated with great caution. Both lines are well marked in all marine service engine rooms approved by the United States Coast Guard. They are usually equally well marked in foreign flag vessels, but the markings may not be in English.

CARGO

Anything that can be moved can be cargo for one or more of the many modes of transportation. Many materials, because they are chemically inert, are unlikely hazards. Other substances, because of their ease of chemical activity or instability, are substantial hazards. They may be shipped in bulk or in small packages, but the characteristic that defines them as hazardous is their chemical stability in any given situation.

These materials, as pure or nearly pure compounds, have been treated individually as hazardous liquids, solids, or gases. There are, however, many other combinations of materials that can be potential hazards in fire situations.

EXPLOSIVES

One of the most hazardous cargos that fire safety personnel can be exposed to is explosives in one or more of their varied forrms. There are bulk limits to the amount that may be legally carried according to the United States Department of Transportation, but even the legal limits can be the cause of a disaster.

The DOT classifications for explosives are:

Class A Explosives: detonating types or otherwise maximum hazards
Class B Explosives: flammable hazards
Class C Explosives: minimum hazards
Oxidizing material: substances that yield oxygen with ease to stimulate the combustion of organic matter

The Institute of Makers of Explosives in their Publication Number 12, "Glossary of Industry Terms," published in October 1978, defines various explosive terms as follows:

Blasting agent: Any material or mixture consisting of fuel and oxidizer, intended for blasting, not otherwise defined as an explosive: Provided, that the finished product, as mixed for use or shipment, cannot be detonated by means of a numbered 8 Test Blasting Cap when unconfined.

Detonating cord: A flexible cord containing a center cord of high explosive and used to detonate other explosives.

Detonator: Any device containing a detonating charge that is used for initiating detonation in an explosive; the term includes, but is not limited to, electric blasting caps of instantaneous and delay types, blasting caps for use with safety fuses, detonating cord delay connectors, and nonelectric instantaneous delay blasting caps.

Explosive materials: These include explosives, blasting agents and detonators. The term includes, but is not limited to, dynamite and other high explosives, slurries and water gels, blasting agents, black powder, pellet powder, initiating explosives, detonators, safety fuses, squibs, detonating cord, igniter cord and igniters.

Explosives: Any chemical compound, mixture or device, the primary or common purpose of which is to function by explosion.

High explosives: Explosives which are characterized by a very high rate of reaction, high pressure development, and the presence of a detonation wave in the explosive.

Low explosive: Explosives which are characterized by deflagration or a low rate of reaction and the development of low pressures.

Mass detonation: Explosive materials mass detonate (mass explode) when a unit or any part of a larger quantity of explosive material explodes and causes all or a substantial part of the remaining material to detonate or explode.

Nitro carbo nitrate: Any blasting agent which has been classified as nitro carbo nitrate under the U.S. Department of Transportation Regulations, and which is shipped in compliance with regulations of the U.S. Department of Transportation.

Primer: A unit, package or cartridge of explosives used to initiate other explosives or blasting agents, and which contains: (1) A detonator; or (2) Detonator cord to which is attached a detonator designed to initiate the detonating cord, which is inserted or attached at the time of use.

Propellant: An explosive that normally functions by deflagration and used for propulsion purposes. It may be a Class A or Class B explosive, depending on its susceptibility to detonation.

Pyrotechnics: Any combustible or explosive compositions or manufactured articles designed and prepared for the purpose of producing audible or visible effects. Pyrotechnics are commonly referred to as fireworks.

Sensitivity: A physical characteristic of an explosive, classifying its ability to detonate upon receiving an external impulse such as impact, shock, flame, or other influences which can cause explosive decomposition.

Small arms ammunition: Any cartridge for shotgun, rifle, pistol, revolver, and cartridges for propellant-actuated power devices and industrial guns. Military-type ammunition containing explosive bursting charges or any incendiary, tracer, spotting, or pyrotechnic projectile is excluded from this definition.

Small arms ammunition primers: Small percussion-sensitive explosive charges encased in a cap and used to ignite propellant powder.

Smokeless propellants (Also called smokeless powder): Solid propellant, commonly called smokeless powders in the trade, used in small arms ammunition, cannon, rockets, propellant-actuated power devices, etc.

Water gels: These comprise a wide variety of materials used for blasting. As manufactured, they have varying degrees of sensitivity to initiation. They usually contain substantial proportions of water ammonium nitrate, some of which is in solution in the water. Some water gels are sensitized by a material classified as an explosive. Some contain no ingredient classified as an explosive but may be sensitized with metals such as aluminum or with other fuels. Under the regulations of the U.S. Department of Transportation, Water gels may be classified as Explosives Class A, Explosives Class B, or oxidizing materials.

MILITARY EXPLOSIVES

A cargo of great potential for disaster is that of military explosives. These cargoes may be found in military aircraft, railroad cars, highway carriers, and military vehicles of all types. Because of the individual packaging of these explosives, which are designed to detonate and destroy life and property, they present extremely hazardous potentials in enveloping fires.

These cargoes may consist of explosives (bombs, rockets, torpedoes, etc.), pyrotechnics (specialized), small arms ammunition, and/or cannon and mortar ammunition.

Bombs

Because of the varied end-uses for bombs dropped from aircraft, there are many types and sizes, ranging in weight from 2 to 5000 pounds. These bombs are classified according to use.

Type of Bomb	Examples
explosive	general purpose (GP); armor-piercing (AP); semiarmor piercing (SAP); light case (LC); fragmentation; depth
chemical	gas; smoke
incendiary	delayed action; immediate action
rockets	same as explosive types
inert types	drill; practice
miscellaneous	

General purpose: The amount of explosive in a general purpose bomb averages about 50 percent of the 100 to 2000 pound (45 to 907 kg) weight for GP bombs. The metal case is designed to withstand impact with ordinary materials even when dropping from flying altitudes. This very fact makes it more dangerous to firefighting personnel. All explosive type bombs, GP, AP, SAP, LC, fragmentation, and depth, can be detonated by an enveloping fire. In situations where the heat is intense enough, they have been known to explode in as short a time as two minutes. In an explosion the strong metal case produces shrapnel that can be lethal. Any piece of shrapnel possessing 60 foot-pounds or more of kinetic energy is considered to be a hazard to life.

Armor-piercing: Made to go through deck armor, heavy concrete structures, and armor-plated vehicles, the nose of armor-piercing bombs is solid and very hard. Explosive content is only 10 percent to 20 percent by weight. All have a delay-type fuse in the tail.

Semiarmor piercing: Very much like the GP, the semiarmor piercing bomb has a very heavy steel case with a thick, solid nose. The SAP types contain about 30 percent explosive by weight.

Light case: The light case bomb has the general appearance of the GP bomb but has a much thinner and lighter case, containing up to 75 percent by weight, explosive.

Fragmentation: Containing about 15 percent to 20 percent, by weight, of explosive, fragmentation bombs are designed to produce the maximum number of fragments traveling at high, lethal velocities. The fragments are designed to fly at right angles to the body of the bomb and so to inflict maximum damage. These bombs also come in clusters which are made to be dropped as a single unit that disperses during the drop.

Depth: Depth bombs are special purpose light case bombs averaging about 70 percent to 75 percent, by weight, explosives. They are designed to cause detonation shock waves in water but nevertheless will produce lethal shrapnel in all cases when exploded out of the water by an enveloping fire.

Identification: For all bombs except chemical and practice types, the general body color is olive-drab with one-inch color coding bands painted at the nose and tail ends of the body. The color code for the bands is as follows:

yellow: high explosive
purple: incendiary
black: inert or practice

Chemical Bombs

Chemical bombs contain materials that under the proper conditions produce, in some cases, physiological and, in others, toxic effects. They may also be merely smoke-producing bombs used to screen or mark troop movements.

The force used to actuate these systems is an explosive, sometimes called a burster. This activator normally extends the full length of the chemical cavity. The hazard of this type of bomb is more in its dispersal of toxic or irritating gases than from high velocity fragments from an exploding case. There is always the possibility of this latter occurring, however.

Identification: Chemical bombs are marked as follows:

Body Color	Band Color	Number of Bands	Markings	Identity
gray	green	2	green	casualty gas—persistent
gray	green	1	green	casualty gas—nonpersistent
gray	red	2	red	harassing gas—persistent
gray	red	1	red	harassing gas—nonpersistent
gray	yellow	1	yellow	smoke

Incendiary Bombs

Made to start fires, incendiary bombs contain jellied gasoline (sodium palmitate + gasoline), phosphorus, thermite, or magnesium powders. The case is usually made of magnesium so that it too can burn and produce more incendiary effect. They contain no explosive as such, but do contain a fuse for starting the incendiary action.

Identification: Incendiary bombs have a gray body with one purple band and markings in purple. The gray body indicates that it belongs to the chemical class of bombs rather than the explosive group.

Pyrotechnic Bombs

Photoflash bombs contain highly flammable materials called pyrotechnics because they produce their brilliant light by means of combustion. They can produce extremely high temperature flames that persist and must be treated according to the procedures for burning metal.

Identification: Pyrotechnics are not marked according to any color coding. In general they are painted gray with markings in black. If they have incendiary as well as photoflash usage, their markings are purple. Quite often, if the body is fabricated from aluminum or magnesium, it will not be painted at all.

Rockets

A military rocket is a projectile, carrying bomb materials, that is propelled by the reaction of burning gases from a propellant. The propellant may be liquid or solid fuel, although very few liquid rockets are found in military cargo. The area where the fuel burns is called the motor and in fact the solid propellant itself, in cylindrical form, is also called the motor. Because the rocket needs no confinement, as do gun-launched projectiles, to produce its velocity, ignition of the rocket motor is relatively easy in an enveloping fire. The autoignition temperature of most solid rocket motors is about 550 °F (288 °C). When this occurs in an uncontrolled situation, the results can be devastating. If the solid fuel has been cracked in an accident, ignition will usually lead to an explosion due to uncontrolled burning caused by the cracks. If the motor remains intact, its ignition by enveloping flames can cause the rocket to be launched with the resultant long-range hazards, the least of which may be widespread fire. In either case, the business end of the rocket, the explosives, is always present and ready to produce the damage for which it was designed. Fortunately all of these devices have safety fuse wires which are supposed to prevent inadvertent ignition, although in an enveloping fire this device may not protect anyone.

Identification: Rockets are marked in accordance with the following color code:

Body	Markings	Identity
olive drab	yellow	high explosive
red	black	low explosive
gray	white (plus one white band)	illuminating

Miscellaneous

There are many and varied types of bombs used by the military for such purposes as signaling, troop marking and point marking. All contain some sort of incendiary or explosive mixture. As such, all constitute a potential hazard in an enveloping fire.

Small Arms Ammunition

According to military definitions, any ammunition up to but not including 20 mm in size is considered small arms ammunition. Military cargo carrying this label may be any caliber 0.60 inches or less. Whether the cargo is to be used for military or for sporting use, it does not contain enough high explosive to do anything but pop like a good-sized firecracker. Modern smokeless powder, the propellant for modern military and sporting ammunition, does not detonate and only does its job of developing high pressures when confined within a gun chamber. The only primary explosive in modern ammunition is a very small amount in the primer. It is this device, in the head of the cartridge, which, when struck with the firing pin of the gun, initiates the burning of the powder and results in the projection of the bullet. Unconfined, in an enveloping fire, the ammunition will pop and throw small fragments, as well as the small primer cups, in all directions. These fragments and metal cups are relatively light in weight and do not possess enough energy to cause injury to fire personnel who are wearing normal plastic face masks and protective clothing—*unless* they are within 50 feet (4.5 metres) of the burning ammunition. Therefore do *not* fight such a fire without protective clothing and masks. Even with protective clothing and masks, fight it from a distance of 50 feet (4.5 metres) or more.

A very real hazard with military ammunition is that of the tracer type spreading the fire as its cases pop. The small amount of incendiary composition in the tail of each bullet is quite capable of starting fires at any distance up to 175 feet (53 metres) from the original fire.

Identification: With only one exception small arms ammunition cases, made of brass, are not color coded, but the tips of the bullets are, as follows:

Tip Color	Identity
none	standard ball (full metal jacket)
black	armor-piercing
silver (aluminum)	armor-piercing incendiary
red (aluminum band)	armor-piercing incendiary tracer
blue (various shades)	incendiary
red (various shades)	tracer
orange	tracer
white (with green band)	frangible
green	experimental

The one exception to the color coding rule for military ammunition is that of high pressure test ammunition. Used for the high pressure testing (called proof testing) of guns only, these cases are a silver color with the bullet having no tip color.

RADIOACTIVE CARGO

Radioisotopes have special transportation hazards, thoroughly discussed in Chapter 13.

SPECIFIC TRANSPORTATION MODES

AIRCRAFT

Fire aboard an aircraft can be extremely hazardous even when the aircraft is on the ground. Because in-flight fire situations are not usually the concern of fire personnel but are fought by mechanical, in-place systems, we shall confine ourselves to aircraft fires which occur on or can be fought on the ground. It is rare that an aircraft fire does not involve hazards to human life so the situations are all the more critical.

Considering the requirements for fire aboard an aircraft we have the following:

Fuel Sources	Oxidizer Sources	Ignition Sources
engine fuel	atmospheric	crash-generated sparks
cabin materials and	on-board oxygen	electrical system sparks
decorations	cargo	hot engine components
metal components		smoking materials
a. magnesium		frictional sparks
b. magnesium alloys		brake overheating
hydraulic systems		cargo
cargo		

Fuel

The hazard of any aircraft fuel resides in its relative ease of oxidation and the amount of energy that is thereby developed. In fact these are the very reasons an aircraft fuel is an aircraft fuel.

The three basic classes of aircraft fuels are gasoline, kerosine, and blends of gasoline and kerosine.

The common root from which gasoline and kerosine are derived is crude petroleum, the same root stock of many of the things we use in order to live.

Gasoline and kerosine have been covered in detail under the classification of hazardous liquids in Chapter 6. However, aircraft fuels, especially the commonly used jet types, are unique blends and therefore have their own characteristics.

	Fuel Type		
Property	Kerosine Base (JET-A, etc.)	Blends (JET-B/JP-4)	Gasoline Base
boiling point	135 °F (57 °C)	325 °F (163 °C)	110 °F (43 °C)
flash point	120 °F (49 °C)	10 °F (−12 °C)	−50 °F (−46 °C)
combustion midpoint ratio	3%	5%	5%
autoignition range	440 °–480 °F (227 °–249 °C)	460 °–480 °F (238 °–249 °C)	800 °–950 °F (427 °–510 °C)

Cabin Decorations

More and more cabin decorations are now made from plastics, which in fire situations present their own particular hazards. These plastics, all of the more modern types, are no longer thought of as having explosive properties as did their predecessors. The early plastics had burning rates that approached explosive limits. These dangerous compositions are no longer legal for public transportation vehicles. Nevertheless, all plastics are organic materials and as such are combustible. When the fire is hot enough they will act as a fuel.

There are two basic types of plastics used for decorations, thermoplastic and thermosetting. The thermoplastic types melt and tend to flow when heated to their melting points. Because these melting points are very high, when flowing they may, in certain circumstances, drip like a burning candle and so transmit the fire. They may also transmit the flames by flowing like lava. In both these cases however, if their combustion point is high enough, these molten globs or flowing streams will not sustain combustion and so go out almost as soon as they leave the immediate flame area. In these cases, rather than contribute to the fuel total, they can be considered to be removing fuel from the original fire, and so are assisting in cooling it or at least in shortening its life.

Thermosetting plastics tend to retain their shape when heated. Some of these thermosetting types will support combustion; others will not. As with many solids, plastics may sublime to form gases when heated to a temperature just below their combustion point. These gases may, in some cases, be combustible, but vary widely in their ease of ignition and the rate of combustion. They may also be toxic.

Because the total mass of decorative polymers in any aircraft is relatively small compared to the rest of the potential fuel mass, the real hazards from burning plastics are not so much in the heat of combustion as from the products of combustion. In general the normal products of organic combustion, carbon monoxide, carbon dioxide, and water, are evolved. In addition, solid particulate, in the form of smoke, will be present due to imperfect combustion. Because other compounds are mixed with plastics to produce desired physical characteristics, there will be additional gases produced that are specific to the type of plastic, composition of the polymer, and the amount of heat impinging on that polymer, as well as the total amount of oxygen available for that specific combustion process.

According to the NFPA the primary cause of fatalities in major fires has been the inhalation of heated, oxygen-deficient air and/or toxic gases. Because polymers are essentially very long chain carbon compounds, the major combustion product will be carbon monoxide. The carbon monoxide usually reacts with other gases to increase the total toxicity. Many polymers contain polyvinyl-chloride (PVC), which, in a fire, will react with the water product of the reaction to form hydrochloric acid. This is a very irritating gas but fortunately, because of that very irritation, gives adequate warning of its presence.

Metal Components

In the minds of many people, the one class of solids that is immune to fire is metals. Nothing could be further from fact. Nearly all metals can react with air rapidly enough to produce flame so long as the temperature is high enough and the surface area of the material is large enough (finely divided particles). Even iron can be made to burn in an oxygen atmosphere when it is in the form of steel wool.

Aircraft designers have a particularly difficult problem concerned with the weight to payload to lift ratios. That is, for any given aircraft design, there will be a given amount of weight that can be lifted off the ground. The more weight that goes into the frame, body and engines, the less payload can be lifted. Aircraft designers have therefore sought out and adopted the lightest, strongest metals available. Cost is given very little if any consideration. The metals that best meet the requirements of aircraft designers, therefore, are the alloys of magnesium and titanium.

The magnesium alloys are often compounded with aluminum in varying percentages for various end uses. For hard, tough end products, the percentages of aluminum are kept as small as 0.5 percent. For light, strong end products that are extremely resistant to fatigue, the aluminum content may go as high as 8 percent to 9 percent with other metals alloyed into the composition. For example, Dowmetal has the following basic composition: magnesium, 87.9 percent; copper, 2.0 percent; cadmium, 1.0 percent; zinc, 0.5 percent; manganese, 0.2

percent; and aluminum, 8.4 percent. In addition to being light, strong and extremely resistant to fatigue, this material also has excellent tensile strength. These properties make it ideal material for aircraft engine parts as well as for aircraft frame components.

However, magnesium is one of the alkali metals and, as such, possesses the family trait of very high chemical reactivity. Magnesium will not react with water at room temperature as does sodium and potassium but it will react with both atmospheric oxygen and nitrogen thus:

$$2Mg + O_2 \longrightarrow 2MgO \ \text{(solid)}$$
$$3Mg + N_2 \longrightarrow Mg_3N_2 \ \text{(solid)}$$

While the ignition point of a pure magnesium ingot is about 1100 °F (593 °C), alloys of magnesium have been ignited at about 800 °F (427 °C). Each alloy, each physical structure (surface area influence) will have its own ignition temperature. In addition, the amount of air flow that the component is exposed to will influence the point of combustion. Because of its high ignition temperature, magnesium components do not usually contribute to the initial fire hazard but will make fighting the fire very difficult if that fire persists and the temperature of the magnesium allowed to rise.

Another popular metal for aircraft construction is titanium, used in both the pure and alloyed form in castings and rolled and formed sheets. In its alloyed form it is very strong, has a high tensile strength, and yet is low in density (lightweight).

Titanium is the tenth most abundant metal found in nature usually appearing as titanium dioxide or **rutile**. Rutile is often sold as a substitute for diamonds in jewelry. The abundance of this oxide, TiO_2, in nature is a clue to the activity of titanium in the presence of atmospheric oxygen:

$$Ti + O_2 \longrightarrow TiO_2$$

Fortunately, this reaction has occurred slowly since the beginning of time and of course is not a combustion reaction, but a redox reaction like the rusting of iron. Titanium forms very strong bonds with oxygen but not at normal temperature, nor with a rapid evolution of energy.

Chemically, titanium is considered to be the head ot the titanium family. This distinguished group includes such members as zirconium and hafnium, all extremely combustible metals, although titanium is not as combustible as its relatives. Nevertheless, turbine components formed from titanium have been reported to catch fire. Once this oxidation-reduction reaction does start, it is almost impossible to extinguish unless the physical situation is such that complete oxidizer removal is possible.

Hydraulic Systems

While continued research has pushed the flammability of hydraulic system liquids toward safer and safer limits, these materials are nonetheless organic compounds and as such will oxidize rapidly. The major hazard is not merely the inherent activity of the oil or synthetic fluid, but that in an aircraft crash or fire, high pressure (2000 to 3000 psi, 14 to 21 MPa) hydraulic lines may rupture and spray the liquid onto hot surfaces or even into an area where an electrical spark may occur. Any spray, because it is made up of fine droplets, has a tremendously increased surface area and so is that much more susceptible to rapid oxidation.

Cargo

Due to the varied nature and the myriad possibilities, just what constitutes hazardous cargo in an aircraft accident is a vast subject. Ease of ignition, toxic gases released either as a direct cargo component or the product of a chemical reaction due to the accident or fire, exposure to corrosive liquids and gases, and even radiation exposure are but a few of the possibilities that can develop from an aircraft accident. For these reasons, the United States Department of Transportation, Office of Hazardous Materials, has issued a set of regulations on what may and may not be shipped aboard aircraft. This list is further defined as to what may be shipped aboard cargo-only carriers as opposed to passenger aircraft. This list is published as part of the Code of Federal Regulations (CFR) Title 49, Part 103, Transportation, Federal Aviation Regulations.

The United Nations and the International Air Transport Association have also developed rules that most countries recognize. These rules are basic but are designed at least to warn by package labeling as to what is in the package. They also prohibit some materials.

Unfortunately the problem of enforcing the regulations on an international basis is of such magnitude that no one group can be responsible. Compliance is left up to the honesty of the shipper. Infractions, intentional or accidental, can result in serious hazards in the event of an aircraft fire or accident. About all the firefighter can do is determine what materials are in a given cargo and, being aware of all possibilities, act accordingly.

Military Aircraft

Modern military fighter and bomber aircraft have been described as an explosive launching platform built around a jet engine, a cynical description that both defines and implies the hazards that can develop in case of fire. Fully loaded with bombs, ammunition, and fuel, such an aircraft might be described as a hazard looking for a place to happen. Extraordinary safety precautions are observed as a matter of course, but fires still occur.

At least seven combustible material possibilities can exist on any military aircraft, no matter what its mission: engine fuel, explosives (bombs, rockets, torpedoes, etc.), pyrotechnics, small arms ammunition, combustible metals, combustible polymers, and hydraulic fluids. All these potential hazards have been covered in detail earlier in this chapter.

Private Aircraft

Engines in private aircraft are still heavily oriented toward the internal combustion engine, although corporate aircraft are fast becoming turbine and pure jet oriented. The internal combustion engine, of course, uses a highly flammable fuel, gasoline, while the turbine and jet use kerosine blends. .

Aside from the difference in size and amount of fuel aboard, the private aircraft fire offers the same potential hazards as commercial aircraft.

It would be foolhardy for fire personnel to assume that a given private aircraft in an accident situation does not carry hazardous cargo. What such aircraft are allowed to carry and what is actually on board that plane can be quite different. In most situations there should be nothing to cause alarm, but, it only takes one hazard to be lethal. Act accordingly.

MARINE

Fire in a marine environment is a situation that everyone dreads. If the vessel is at sea the safety and very existence of the vessel and crew are jeopardized. If the vessel is at or near a berth, not only is its own safety in question but the surrounding area can be endangered by either fire, explosion, or toxic gases. The vessel's isolation must be considered in the fire-fighting strategy. Ironically, water, one of the best extinguishing agents known, is the cause of that isolation, which more often than not works against, rather than for, crew safety and fire extinguishment.

With the widespread use of flammable substances, the ability of a marine cargo to catch fire is more a matter of degree than a division into flammable and nonflammable materials. The causes may be categorized as follows:

Fuel Sources	Oxidizer Sources	Ignition Sources
decorative materials: paint, plastics, wood, cloth,	atmospheric cargo	boiler flame crash-generated sparks electrical system sparks seawater/electrical system interaction
engine fuel cargo		cargo hot engine components

Every commercial vessel is a business venture and as such is designed to make money for its owners by carrying its cargo in the most efficient manner possible. There are, therefore, a number of different types of vessels each with its own specific type of potential hazard as well as the hazards general to all marine carriers.

The basic types of vessels are:

Tankers: Designed to carry both liquid and gas cargoes, they carry some of the most dangerous materials known. In spite of this, this type of vessel is being made larger and larger, e.g., supertankers.

Refrigerator ships: Used for long haul, bulk transportation of perishable substances, these carry a potential hazard in their refrigeration systems.

Ore boats: These are specifically designed for bulk transportation of solids in just about any size, from large aggregate to finely ground powder. The hazard with these vessels can be with the material and the particle size if it is finely divided.

Liner service vessels: These are designed to carry general cargo in different holds, or on the deck, over well-defined routes. The potential hazards here are those of cargo and marine hazards in general.

Passenger vessels: Designed to carry people in varying degrees of comfort, they have an abundance of flammable decorative materials throughout the vessel. They may also carry some cargo.

Tramp vessels: Made to carry any cargo, any time and to any place, they are not always the best maintained and so can be a great potential hazard in fire and accident situations.

Decorations

As with aircraft decorations, more and more plastics are being used in marine design, and the same hazards are present in marine fires as in aircraft fires.

Another large source of potential hazards in marine fires is the paint used both to protect and beautify the wood, iron, or steel of the vessel. Fighting a fire in an enclosed space is hazardous enough in itself, but when a source of toxic fumes is also burning the hazard becomes much greater. The source of lead fumes can be paint.

Paint is a chemical mixture of pigment for color and sometimes for flow characteristics, vehicle that makes it spreadable and later polymerizes to form the protective shield, and solvent which keeps the vehicle is a spreadable condition.

While prime pigments, such as red lead, white lead, and zinc chromate, help prevent rust and corrosion of metal, they can, under fire conditions, vaporize and provide an extensive hazard to fire-fighting personnel. Pure lead melts at about 620 °F (327 °C), but it vaporizes at 1350 °F (732 °C). Finely divided lead particles can be carried good distances by the smoke from burning lead type paints. Lead poisoning is cumulative, so that small doses over a long period of time will eventually lead to serious results if countermeasures are not taken. Proper medication administered by a physician should be started as soon as lead poisoning has been diagnosed.

Ignition Sources

When two steel vessels, each weighing thousands of tons and moving even at slow speeds, collide with each other, the sudden transformation of all that kinetic energy results in sparks and hot metal as uncontrolled ignition sources. With the other damage that is bound to occur, such as ruptured fuel lines or even tanks, all of the conditions for fire are present:

Electrical: In a seawater environment, dissolved salts are ionized and therefore will conduct electricity very well. By actual analysis seawater has been shown to contain essentially all metallic and nonmetallic ions normally found in nature.

The major components in seawater are: sodium, 1.1 percent; chlorine, 2.0 percent; magnesium, 0.13 percent; calcium, 0.04 percent; and potassium, 0.04.

Ruptured electrical circuits, in contact with a conductive pool of seawater, can lead to overloaded circuits; overloaded, and therefore hot, wires; and sparks. All are excellent sources of ignition for cargo, engine fuel, or decorative flammables.

Hot engine: As long as engines (or turbines) are less than 100 percent efficient, there will be unused energy in the form of heat in and around the engine room. Any flammable spill or ruptured line can provide the rest of the components for the hazard.

Marine Refrigeration

The potential hazards on a refrigeration ship are all of those found on a regular commercial vessel, plus that of the refrigeration equipment. The hazard of the refrigeration equipment is not so much that of the reduced temperature as it is that of the machinery and the chemical compounds that are used to achieve lowered temperatures.

Depending on the size of the area to be cooled there are three basic methods of refrigeration: compression, absorption, and steam-jet.

Any mechanical cooling system works on the principle that a liquid must absorb heat as it evaporates. The cooling effect produced by an electric fan blowing in your face is no different. The evaporation of water from the surface of your skin removes heat from the body and so produces a cooling effect. In areas where the relative humidity is very low, as in the desert, evaporation can be very rapid and the cooling quite pronounced.

The compression and also the absorption systems produce the lowered temperature by a closed cycle wherein a liquid is evaporated to a gas followed by condensation back to a liquid, again and again. This is called the refrigeration cycle.

In the compression system, a compressor changes a chemical compound that is a gas at room temperature into a liquid by increasing the pressure about 100 psi (698 kPa). When the liquid flows into the area to be cooled, it is allowed to evaporate under reduced pressure, and as it evaporates it cools the coils in which it is flowing. The coils then absorb heat from the surrounding area. The gas now returns to the compressor to be changed back into a liquid and the cycle repeated.

In the absorption system, the cooling compound is normally a liquid at room temperature and is changed to a gas containing a great many liquid droplets. The change is accomplished by heating the liquid by flame. As the gas-liquid mixture evaporates further, it cools the desired area by means of coils. This system has no moving, mechanical components.

The steam-jet system uses water in two forms, as the coolant and as the producer of evaporation. This system works on the basic principle that the lower surrounding pressure lowers the boiling point of a given liquid, in this case water. In addition the lower boiling point speeds the evaporation process, and therefore the cooling.

The heart of the steam-jet system is a chamber wherein a high velocity jet of steam can be passed across a volume of water. As it passes across the water, the steam carries with it gas molecules that exist above the liquid. This reduces the pressure in the chamber, and so at the surface of the liquid, evaporation takes place. This evaporation action absorbs heat from the body of the liquid that still remains. The cooled water may then be pumped through coils in the area to be refrigerated.

Because of its simplicity and economy of operation this system is very popular for large industrial and marine refrigeration installations.

Refrigeration Hazards

The hazards of refrigerator systems lie mainly in the compounds that undergo the change of state from liquid to gas and back again during the refrigeration cycle. In the steam-jet system, the refrigeration medium is water so that very little hazard exists from it. However, this same system does make use of high velocity steam and this can do great damage should the human body come in contact with it because of a steam leak or burst pipe—not uncommon in an accident.

In absorption and compression systems, ammonia is a very popular liquid-gas refrigerant. This ammonia, however, must not be confused with household ammonia, which is a small percentage of ammonia gas dissolved in water, becoming ammonium hydroxide, which is a base.

$$NH_3 + H_2O \longrightarrow NH_4OH$$

Refrigeration ammonia is known as **anhydrous ammonia**, which has very different characteristics. The word anhydrous comes from the Greek and means "without water". The opposite of anhydrous is hydrated, which indicates that a given compound has one or more water molecules attached to its molecule.

Anhydrous ammonia is a colorless gas that is very irritating to the throat, lungs, skin, and eyes. Odd as it may seem, anhydrous ammonia is flammable at concentrations of 15 to 17 percent in air. Because of its high solubility in water, potentially explosive mixtures of this gas can be turned into harmless, chemically basic solutions by just spraying fog nozzle water into the area.

Other than its flammability, the ammonia molecule is more like the water molecule than any other compound known at this time. Water has a high association constant and is thought to be constructed by the joining together of the molecules into long tetrahedrons. Ammonia has a similar structure. Both ammonia and water are colorless liquids, and ammonia readily dissolves most of the compounds that dissolve in water.

Anhydrous ammonia, in itself, is not really an active compound. It becomes very active, however, with even the slightest bit of moisture. This is one of the reasons it is extremely hazardous to human life. It reacts quickly with the moisture of the body to form a strong basic solution, which then destroys surrounding tissue.

RAILROADS

Unlike marine fires, railroad accidents, fires, and the resulting hazards are not isolated by water unless, of course, the accident occurs on a bridge. The lack of isolation can be a positive factor in extinguishing any fire but it can also be a very negative factor, making the hazard potential even greater. In populated areas, toxic gases, fire, or explosions can and often have led to disasters.

The following combustion or hazard requirements may be present in railway accidents:

Fuel Sources	Oxidizer Sources	Ignition Sources
engine fuel	atmospheric	crash-generated sparks
cargo	cargo	electrical system sparks
		hot engine components
		brake overheating
		cargo

Of all the hazard potentials in railway accidents, the greatest seems to be the cargo. Rupture or prolonged heating of a bulk carrying car, especially for liquids or gases, can lead to toxic fumes engulfing an area, or flammable liquids coming in contact with one or more of the ignition sources listed above. The result has been recurrent disasters in sufficient numbers to jolt the Materials Transportation Bureau of the Department of Transportation into issuing new rules for uninsulated pressure tank cars for the transportation of hazardous materials (Amendment numbers 173-108, 179-19).

On February 20, 1977, in Dallas, Texas, an Atcheson, Topeka and Santa Fe Railway freight train derailed. In the derailment a tank car loaded with over 32,000 gallons of propane sustained a tank-head puncture near the base of the head. The escaping propane ignited and the resulting torchlike flame impinged upon and heated another tank car containing over 30,000 gallons of isobutane. After about 40 minutes of flame contact the tank car carrying the isobutane exploded with violence. The tank separated into three major and several minor parts. There were no injuries but the loss was well over 3.5 million dollars.

On November 9, 1977, a railway derailment in Florida resulted in the overturning of a tank car carrying about 33,000 gallons of anhydrous ammonia. According to the United Press International news service, the car split open pouring the liquid out onto the ground where it turned into a gas and spread through a residential community about 4.5 miles from downtown Pensacola. The UPI reported one man killed and 16 others hospitalized. It was also reported that this was the nineteenth train derailment in 18 months and the second serious derailment involving ammonia-carrying tank cars in the general area in less than one month.

Because of their fluid characteristics, liquids and gases present the greatest hazards as cargo. However, these materials are not the only potential hazards in railway accidents. Bulk freight cars transporting solids have produced violent explosions.

According to an Associated Press report, on Friday, November 11, 1977, thirty tons of dynamite aboard a freight train in the freight yards of the city of Iri, South Korea, exploded killing about 60 persons and injuring about 1300 others. The explosion was heard more than ten miles away and dug a crater about 60 feet deep. According to AP reports, this disaster may have been caused by a watchman's candle setting fire to his sleeping bag which was in one of the freight cars. This is most certainly an unusual ignition source as well as an abnormal place to sleep—surrounded by dynamite.

The watchman escaped injury because he ran as far as he could from the area as soon as he found he could not put out the fire in the sleeping bag.

Potentially explosive cargo fires should not be fought. The best course is to evacuate the entire area for at least 2000 feet (609 metres), about 7 football fields, in all directions. If the DOT labels proclaim a shipment to be explosive, *don't fight it—evacuate.*

When the hazard is from a leaking toxic gas such as ammonia, 2000 feet (609 metres) may not be enough, especially downwind. Remember winds shift and a safe area may suddenly become the exposed area.

HIGHWAY

While marine fires tend to have some degree of isolation due to their natural surroundings, as do railroad fires to a lesser degree, highway fires involving hazardous materials possess the greatest potential for disaster. They can and often do occur right in the middle of populated areas.

Basically automotive transporters have the following fuel, oxidizer and ignition sources:

Fuel Sources	Oxidizer Sources	Ignition Sources
engine fuel	atmosphere	crash-generated sparks
tires	cargo	electrical system sparks
body (if wood)		hot engine components
cargo		brake overheating
		personnel carelessness
		tire fires
		cargo

Except for the control afforded by the rails, trucks are quite similar to railroad cars. That exception leads to a large increase in the statistical risk when carrying hazardous materials. Furthermore, while there is only one engine on a train of cars, each truck or, at most, two trailers have a hot engine and fuel with all the resultant chances for ignition.

It would seem from a study of case histories that truck, tractor, and trailer tires are one of the greatest sources of highway transportation fires. One of the problems is that these fires can get started just from over-the-road friction if the tires are not properly maintained and kept under observation. A case has been reported where a hot tire, left to cool in a rest stop, continued to heat up until visible fire broke out and ignited the body, finally destroying the cargo. Fortunately this cargo did not contain hazardous materials.

When fighting truck or trailer fires, the driver is the best consultant fire personnel can have. Responsible drivers not only know what they are carrying, but know how best to handle dangerous cargo. They therefore should be your first contact, if possible.

Once the fire spreads to the truck or trailer body, the cargo will soon become involved. Where hazardous materials are concerned, you must decide, with the help of the driver, the relative hazard risk. If there is any chance of an explosion, *do not hesitate,* but clear the area of traffic at once, evacuate all buildings, and move everyone back at least 2000 feet (609 metres).

In a small town near the eastern seacoast, a collision between a small car and a trailer rig carrying explosives caused the gas tank on the truck to catch fire. The driver, realizing the danger, began moving everyone back the recommended 2000 feet (609 metres). When the volunteer fire department arrived, they felt they knew more about fighting fires than the driver and refused to listen to his warnings. When the cargo exploded, several firefighters and one spectator were killed, and several vehicles destroyed. The driver and those who listened to him were not injured.

As with railway accidents, when a tank truck is involved and carrying toxic gases, 2000 feet (609 metres) may not be enough. For this situation weather conditions must be taken into account.

Flammable gases, especially those gases with high cohesive forces, can work their way into sewer systems. Once in the sewer network, they can cover a broad area and do widespread damage if ignited.

Many years ago in a midsized western town, a tanker loaded with propane parked on one of the main streets while the driver took a meal break. The truck was well marked as to what it was carrying, but it is believed that a small pickup truck, while trying to park behind the tanker, hit and damaged one of the outlet valves. The driver of the pickup told no one, or perhaps did not know what he had done. The valve leaked enough gas for the propane to get into the sewer drains beneath the tanker and over a period of time, the propane spread out beneath several blocks of the town.

It is not clear what the ignition source was, but something did ignite the propane, and several blocks of that town disappeared: pavements, buildings, and all.

REVIEW QUESTIONS

1. Discuss the relationship between chain length and energy in aromatic compounds.
2. What is heat of combustion? Why is it important? How does it vary with chain length?
3. What is an octane rating? How is it determined?
4. How are fuel oils rated? What does the rating mean to fire personnel?
5. What is a cetane number? How is it determined?
6. What type of coal is the most hazardous? Why?
7. What are class A, class B, and class C explosives? Detail the differences.
8. What is an oxidizing material? How does it differ from an explosive?
9. Discuss in detail the different types of jet fuels. What are the relative hazards?
10. Discuss the difference between thermosetting and thermoplastic materials.
11. What are a few of the specific hazards of aircraft metals?
12. Discuss in detail the potential ignition sources in marine accidents.
13. What ignition sources can be expected in railroad accidents?
14. What is the hazardous material that is often present in large refrigeration systems? Why is it hazardous?
15. Discuss the procedure that should be followed in a potentially explosive fire.
16. What type of fire is most often experienced with trucks?
17. Why are liquid fuels most popular in the transportation industry?
18. Why do the longer chain hydrocarbons have higher heats of combustion?
19. Why are diesel fuels cheaper than gasoline?
20. Define the different types of coal and the dust hazard of each.
21. List and explain the Department of Transportation placards for the transportation of explosives.
22. List the types of bombs used by the military.
23. List the hazards of a fire involving small arms ammunition and the precautions to be taken.
24. List the hazards found in fighting military aircraft fires.
25. What are the hazards of ammonia in refrigeration? Will it burn? How should firefighters handle it?
26. List the sources of fuel, oxidizer, and ignition found in railroad transportation systems.

Extinguishing Agents/ 11

Fire can be extinguished by several different methods. Except for a few special classes of fires which will be discussed later, extinguishment can be resolved into removing one or more of the three sides of the fire triangle. Removal of the oxidizer, removal of the fuel, or removal of the reaction-developed energy (heat) faster than it is being generated, will all lower the temperature and eventually extinguish the fire. It could be argued that even these three methods could be resolved into one: *cool it*. The basis for this argument is that removing either the fuel or the oxidizer or the developed energy reduces the heat output very rapidly and so the entire chemical reaction that is producing the fire undergoes a cooling down.

Removal of the oxidizer is not always possible or practical. In a small fire, where isolation is possible, the oxidizer supply can be cut off so that the fire will self-extinguish by using up the oxygen that remains. It is said then to be smothered. This is exactly how a candlesnuffer works.

In a fire of any magnitude, however, the oxidizer being supplied by the atmosphere cannot be cut off. In this situation the use-it-up technique must be applied to the fuel. The fire is allowed to burn itself out with control exercised by limiting the amount of fuel available from the surrounding area. This is the method followed in those instances where firefighters limit their efforts to protecting exposure as in the case of a fully involved barn with a limited supply of water—they control the fire to the extent of preventing its spread to the house and other buildings.

On a smaller scale, fighting a grass fire with brooms is also removal of the fuel by separating the unburned material from that which is fully involved. Another way of removing the fuel is used in controlling large forest fires where deliberate backfiring is applied. Here the fuel is removed by controlled burning before the main wave of fire advances to that point. When it does arrive at the backfired area, the fuel is gone, one leg of the fire triangle is missing, and the fire goes out.

233

WATER

The removal-of-the-heat approach has always been, and still is, the most common and basic method. The most plentiful liquid on earth is the most effective means for removing the heat from a fire. If a gigantic chemical corporation had set about developing the ideal fire extinguishing agent, they would have undoubtedly, after spending millions of dollars on research, developed hydrogen oxide, or, by its more common name, **water**. History shows that the vast majority of all the scientific discoveries and developments relating to fire suppression center around the improvements in the means of delivering water to the fire. From the bucket brigade, hand pumpers, and early steamers, through the era of gasoline-powered rigs to the most modern diesel units, there has been a continual improvement in pumps, hose, nozzles, and water systems in general.

While the cost of water, its availability, and physical and chemical characteristics are just about perfect for fire extinguishment, it does have some disadvantages.

Advantages	Disadvantages
• relatively high heat absorption	• relatively high freezing point
• relatively low cost	• low viscosity; runs off quickly
• extremely plentiful	• electrolyte in its natural state
• normally a liquid	• high surface tension
• low viscosity; pumps easily	• contains oxygen; reacts with some substances
• found in nature	• high specific gravity

Of the many thousands in the world, there is no other liquid that could be used as a direct substitute. Other fire extinguishing compounds have been developed to overcome specific disadvantages of water, or for specific fire fighting situations, but there is no known substance that, even if it were abundantly available, could do the job as well or as economically.

Every school child learns that water is colorless, odorless, and tasteless, but adults know it isn't. Pure water may be colorless, odorless, and tasteless, but water as it is found in nature or provided in water supply systems is not. Furthermore, such water varies in some of its chemical and physical properties.

Some of the physical properties of pure water are:

boiling point (at 1 atmosphere pressure)	212 °F (100 °C)
freezing point	32 °F (0 °C)
density (at 4 °C)	1.0 g/ml
specific heat	1.0 cal/g/°C
heat of vaporization	538.6 cal/g

While most commonly available water will vary slightly on either side of the above values, it will not vary enough to reduce its real effectiveness as an energy remover from combustion reactions.

One of the characteristics that makes water so effective as an energy remover is the large amount of heat required to change it from one physical state to another, such as from a liquid to a gas, or even to raise its temperature within any one of these states. When compared to other substances, especially those that are combustible, the specific heat of water is quite high.

SPECIFIC HEAT

By definition, the **specific heat** of a substance is the amount of energy (in joules) required to raise the temperature of 1 gram of that substance 1 degree centigrade. In order to use a common reference material, water has been chosen as the standard. In other words, the specific heat is a comparison between the amount of energy needed to raise the temperature of 1 gram of water 1 degree centigrade and the amount of energy needed to raise the temperature of 1 gram of the substance 1 degree centigrade. One of the big advantages of water for fire extinguishing is that it requires more energy to raise its temperature than almost any other substance. For combustion reaction energy removal, nothing could be better.

Some specific heats are:

Substance	*Specific Heat (Joules/g/°C)*
water	4.18 (at 20 °C)
wood	1.76 (at 20 °C)
sugar	1.15 (at 20 °C)
charcoal (dry)	0.92 (at 20 °C)

Translated into practical fire-fighting terms, water's high specific heat indicates that when water can be used on a burning substance, the temperature of the substance will go down more quickly than the temperature of water will go up. In energy terms, an equal weight of water will reduce the energy of an equal weight of a substance by a factor equal to the specific heat of the water divided by the specific heat of that substance. For example, in a wood fire one gram of water will cool 4.18/1.76 or 2.37 grams of wood 1 °C or 1.8 °F.

It might be argued that we don't want to cool the fuel, just absorb the energy produced by the chemical reaction. However, the energy that is being produced appears both as flame, or radiated heat, and as an increase in temperature of the fuel, an increase that raises it above its ignition point. Cooling the fuel therefore does absorb the critical energy that is being produced, which is keeping the fire going.

The high specific heat of water is very important in dealing with flammable, water-miscible, solutions such as ethyl alcohol or acetone. Adding water to such solutions not only cools the mixture but also radically changes the specific heat as well as the flash point of the solution. Because liquids do not burn as liquids, but must vaporize first, this cooling down by dilution can be an important method of extinguishing or even preventing fires. A slight miscalculation, however, can result in serious problems. Consider the situation where fire occurs in an uncovered container of alcohol. While the water will cool and dilute the alcohol, it will also add to the volume of liquid in that container. The question then becomes, "Which will occur first, extinguishment or overfill with the potential of a flowing fire?" If it is impossible to assure extinguishment prior to overflow, the more intelligent approach might be to protect exposures until sufficient fuel has been consumed to reduce the level to the point where extinguishment by cooling and dilution can be safely accomplished.

HEAT OF VAPORIZATION

In addition to its relatively high specific heat, water also has a relatively high **heat of vaporization.** This characteristic is most important in fighting fires because the heat of vaporization is defined as the amount of heat a material absorbs in changing its physical state from a liquid to a gas. Water absorbs a tremendous amount of energy when it changes from a liquid to its gaseous form (steam) as it is poured on a fire. As we learned earlier, energy always flows downhill, from the higher energy level to the lower energy level. Water, in absorbing the large amount of energy as it changes from a liquid to steam, therefore offers a big energy hole that drags a great deal of heat away from the fire reaction. The hole is even deeper when a fog nozzle is used, for more water molecules are exposed to the burning substance and they extract more energy as they are changed from a liquid to a gas. Every pint of water will absorb about 1023 kilojoules of energy, or about 8190 kilojoules for every gallon that is vaporized.

WATER ODDITY

A very unusual characteristic of water is the manner in which its volume changes in relation to the temperature change it undergoes. At normal temperatures, such as 68 °F (20 °C), water acts like most other liquids: that is, as the temperature rises, the volume increases. The reverse also holds true: as the temperature is decreased from 68 °F (20 °C), the volume decreases until 39 °F (4 °C) is reached. At 39 °F (4 °C) and below, water begins to act quite differently. It reverses the volume contraction it has been going through as the temperature was lowered and at about 39 °F (4 °C) it begins to expand until it has reached just over 10 percent of its original volume.

The reasons for this unique behavior lie in the molecular structure of water. Actually we are not rigorously correct when, as chemists, we write water as a simple molecule H_2O. Water has been shown, by means of X-ray studies and exotic molecular weight determinations, to be a polymer. These studies have also shown that the amount of polymerization, or really molecular association of water, increases as the temperature decreases. We should, therefore, be writing water as $(H_2O)_x$. Only in the gas phase is water plain old H_2O.

The X-ray studies have also shown that the polymerization or molecular association is due to the presence of hydrogen bonds between water molecules. To vaporize water, as is done when water is used to extinguish a fire, requires breaking these hydrogen bonds. The breaking of hydrogen bonds requires more energy than is normally required to overcome the normal attraction between molecules in other liquids. Thus the relatively high specific heat and heat of vaporization of water.

SURFACE TENSION

Because the above two characteristics are so favorable to fire extinguishment, it is important to provide maximum contact between the burning surface and the water molecules, or to increase the surface contact between the two surfaces. This can be accomplished by lowering the surface tension of the water so that it spreads more evenly, or wets the surface better. **Surface tension** is defined as the force that results from the difference between the molecular forces acting upon the molecules *beneath* the surface of the liquid and those forces acting upon the molecules *on* the surface of the same liquid. Because the molecules beneath the surface are completely surrounded by other molecules, they have a greater total force acting upon them than those at the surface where they are, at least partially, exposed to the atmosphere. The resultant difference is a contracting force, called surface tension, which tends to pull the surface into the minimum area. When there are no enclosing walls for the quantity of liquid, it breaks up into a shape restrained only by that contracting force. The resulting shape is a sphere because a sphere produces the minimum surface area for any given volume. When the surface tension is reduced by additives that reduce the internal molecular forces or any materials (such as dirt) on the surface, the spheres are kept from forming. The water then flows out into a film rather than forming beads. An example of this is the way water flows, wetting a dirty, unwaxed surface on an automobile, as opposed to the way it beads on a newly waxed surface of fire apparatus. The newly waxed surface produces less friction for the water and so allows the internal molecular forces to dominate and form the spheres.

WET WATER

Many chemical additives will make water wetter by lowering its surface tension. Soaps and detergents are two common agents that will do this. Water with such additives blankets the surface and penetrates the material more effectively. A distinct improvement in extinguishment can be observed when wetter water is used on smoldering fires in upholstered furniture and baled cotton, and on woodland fires that have penetrated deeply into the forest duff.

As soon as early experiments with common laundry detergents and soaps proved the merits of wetting agents, many fire protection rigs began carrying these materials, premixed with water, in their booster tanks. It wasn't long before in-line proportioners were added to the plumbing of some engines so that water could be made wetter as needed, by the simple act of opening a valve while pumping a line. Such applications were usually limited to small fires with the wet water pumped through booster line or preconnected 1½-inch lines. These early surface-active agents were found to be very destructive to valves and packings and so have fallen into disfavor.

SLIPPERY WATER

In place of the early surface-active agents a new water additive has become very popular. This is polymerized ethylene oxide, sometimes called "poly". Ethylene oxide itself is a very toxic gas, but when polymerized it is so nontoxic that it can be used safely as a food additive. In its polymerized form it is a white solid that has many and varied uses. Ethylene oxide, as a molecule, has the chemical structure

$$\begin{array}{ccc} & H & H \\ & | & | \\ H- & C-C & -H \\ & \diagdown\diagup \\ & O \end{array}$$

and as a polymer it would be diagrammed:

$$\begin{array}{ccccccccc} & H & H & H & H & H & H & H & H \\ & | & | & | & | & | & | & | & | \\ R- & C-C-C-C-C-C-C-C & -R \\ & \diagdown\diagup & & \diagdown\diagup & & \diagdown\diagup & & \diagdown\diagup \\ & O & & O & & O & & O \end{array}$$

These long, linear, unbranched chain polymers make excellent film-forming compounds. It is this film-forming property that makes polymerized ethylene oxide so effective in reducing turbulence, and therefore energy losses, in pumped lines. It is added as a mixture of finely divided, solid ethylene oxide polymer suspended in water. The chemists' term for this mixture is slurry. The slurry is added to the pumped water at the rate of about one gallon of slurry to about 5000 gallons of pumped water. This rate has been estimated to improve water delivery by about 45 to 48 percent.

The exact mechanism by which these long, linear polymers, such as ethylene oxide, reduce the required energy needed for pumping is unknown. It is known that the turbulence-reducing property is directly proportional to the length of the polymer. Ethylene oxide is said to be two to three times more effective than any other additive thus far developed. It has been calculated that 95 percent of the energy loss in pumped, high velocity stream lines is due to the friction between water molecules themselves, which are believed to be pyramidal-shaped, polymer-like, aggregated molecules. Any very long, unbranched polymer, when mixed with water, is believed to form molecular films over the sharp water pyramid tops and reduce the amount of shear friction as these molecules try to slide by each other.

The effect, without ethylene oxide, is like rubbing two pieces of coarse sandpaper together. If we think of the abrasive granules on the sandpaper as water molecules trying to slide by each other, we get an idea why there is such energy loss in high velocity pumped lines. Adding ethylene oxide slurry to water would be like filling in between the abrasive particles of the sandpaper with a plastic. The sharp edges are coated over and frictional losses considerably reduced.

The practical results of the big reduction in friction is less internal flow turbulence, which results in:

- greater volume output for equal pressure
- equal volume output with reduced line diameter and, therefore, greater mobility with smaller line
- greater nozzle pressure
- greater stream reach
- less stream divergence

According to T. C. Clough, actual tests with the New York City Fire Department showed that a 2½-inch line delivered 250 gpm (945 l/m) without the additive, and with the additive the same amount could be delivered with a 1½-inch line. With the additive, the 2½-inch line delivered almost as much as a 3½-inch line, all of equal length.[1]

These same tests are reported to have demonstrated a large increase in nozzle pressure and over a 25 percent increase in stream reach with far less stream divergence.

Other tests by Union Carbide Corporation, a manufacturer of polymerized ethylene oxide, show that this polymer is absolutely nontoxic, having no adverse effects on plant or marine life; will be equally effective in fresh or salt water; and will become a harmless powder over a period of time when exposed to sunlight.

---

[1]T. C. Clough, "Research on Friction Reducing Agents," *Fire Technology,* vol. 9, no. 1 (February 1973).

THICKENED WATER

While the low viscosity of water can be an advantage in situations where penetration is needed, low viscosity can also be a disadvantage because it causes the water to flow away from the combusting materials and so take with it all its potential cooling effects. The heat that would have been used up to vaporize the water remains to add to the fire problem.

In order to overcome the flow, experiments have been going on for some time with materials that increase the viscosity of water by thickening it. One of the most promising of these thickening agents is a synthetic cellulose gum, sodium carboxymethylcellulose. It is also known as sodium cellulose glycolate, CMC, CM cellulose, or just plain cellulose gum. This is the same material that is used to thicken artificially some milkshakes, where instead of more expensive ice cream, this relatively inexpensive powder is added. Cellulose gum is used extensively in food stuffs and even ice cream itself.

The mechanics of the thickening process are scientifically complex but can be simplified. Once in water the very hygroscopic cellulose gum, acting like millions of microscopic sponges, begins to swell until each molecule has increased in size a thousand or more times. The result is increased intermolecular friction, and the thickened water begins to act like a material somewhere between a liquid and a solid.

In this way the water, which is locked up with the millions of sponges, is held at the scene of the flame. The water is vaporized and cooling of the fire takes place.

When properly mixed, the thickened water will stick to burning material much better than ordinary water. However, cellulose gum is a hydrocarbon, and as such, under heat it will form a film over the materials to which it has been applied. This film can produce a mess in the wrong applications, but it is often better than letting fire destroy the material. Thickened water is ideal for forestry service fire work, where its properties provide a better, more controlled fall and superior blanket when dropped from aircraft.

Thickened water has its disadvantages as well as its advantages. Two of the major disadvantages are:

- Surfaces where thickened water have been applied, or onto which it has splashed, become very slippery and hazardous. This is especially true of metal stairs.
- Where penetration is required, thickened water will not do the job; it prevents penetration.

Other materials beside cellulose gum which have been used to thicken water are: ammonium phosphate, including diammonium phosphate (DAP) and monoammonium phosphate (MAP); ammonium sulfate; and bentonite clay.

While all thickening agents produce non-Newtonian viscosity in varying degrees, bentonite clay, in certain ratios with water, produces classic examples of non-Newtonian flow. The mixture will set up as thick as mayonnaise and yet when physically disturbed, with either a paddle or a pump, thins out to resemble ordinary water flow characteristics. On a surface and mechanically undisturbed it once more sets up and becomes a thick gellike material.

SOME DISADVANTAGES
OF WATER ITSELF

While water is the almost perfect extinguishing agent, it must be remembered that it is just that–*almost*. Water does not counteract all chemical reactions. In fact, with certain elements and compounds, it will promote redox reactions. Fighting fires in such substances therefore requires different weapons, as we shall discuss later.

We don't always understand how water contributes to all of its reactions but we do know:

- In some reactions water seems to assist in the formation of ions, which react more easily than the atoms or molecules from which the ions are formed.
- In other reactions, water seems to act as a bridge which allows the reacting molecules to get together and so to react.
- In still other reactions, water seems to have merely a catalytic effect. Nevertheless it does speed up even these reactions.

The most notable water reactions are with the alkali metals: lithium, sodium, potassium, rubidium, cesium, and francium. In examining the chemical activity of these elements, we find that this fault of the almost perfect extinguishing agent is not really the fault of water at all. The alkali metals are characterized by what could be called hyperactivity: they are all so active they do not exist as elements in contact with the atmosphere or water. Even their mixtures, alloys, and metallic solutions react violently.

Other notable water reactions occur with the families of compounds called nitrides, carbides, hydrides, silicides, sulfides, selenides, and telurides.[2]

Another undesirable characteristic of water, for some types of fire suppression, is its specific gravity. Most combustible liquids and many combustible solids have a lower specific gravity than water and therefore will float on its surface. Pouring water onto such a burning liquid or solid is an excellent way to spread the flames. A story is told of firefighters in a small western town who were washing down a large gasoline spill from a tank truck. Unfortunately the gasoline caught fire. When they tried to drown the fire, they actually floated the fire across the street and into a store. The store caught fire. Realizing what they

---

[2]This "-ide" class of compounds are those in which an element has combined with another element less electronegative (in the oxidation state) than itself.

had done, they thought they saw an easy solution. They went around the back of the store with their lines and washed the fire out through the front door . . . and right across the street into another store! Water is the almost perfect extinguishing agent . . . *almost.*

To be an effective extinguishing agent for a fire involving materials of lower specific gravity, the water must be applied in extremely small droplets and even then with a chemical additive to reduce the surface tension to the extreme. Only then is there even a remote chance of achieving extinguishment of any liquid whose flash point is lower than ambient temperature.

CARBON DIOXIDE

Under the general fire extinguishing principle of "cool it," one method was to remove the oxidizer. This of course is not always possible, especially when the oxidizer is built into the molecule. However, with relatively small fires feeding on atmospheric oxygen, it is quite often an excellent technique.

One compound exists that, because of its ancestry, is an outstanding fire extinguishing tool. Carbon dioxide (CO_2), the common product of combustion reactions involving organic fuel, such as wood, natural gas, and gasoline, is so heavy it will settle around and thus blanket a fire, denying the fuel further access to atmospheric oxygen.

This characteristic of carbon dioxide reminds me of the child of a broken home; CO_2, being born of a combustion reaction, knows exactly how to separate the combatants, the fuel and oxidizer. Unfortunately not enough carbon dioxide is generated in a combustion reaction to make organic fires self-extinguishing.

There is a popular misconception that carbon dioxide in its solid form, also known as dry ice, is effective as an extinguishing agent because it cools the fire very rapidly. While there is, undoubtedly, some cooling effect, it is miniscule in comparison to the oxygen-excluding effect of a blanket of gaseous carbon dioxide.

The red cylinders of carbon dioxide commonly found in portable fire extinguishers and fixed automatic fire suppression devices, and carried on some specialized fire apparatus, contain both gaseous and liquid CO_2 under pressure of about 60 times atmospheric, about 870 psi (6.07 kPa). The pressure serves multiple purposes. A large volume of gas can be contained in a small package at high pressure; the pressure delivers the gas through the discharge hose and delivery horn; and the sudden lowering of the pressure results in a marked lowering of the temperature of the gas with a portion of the discharge becoming solid as the temperature of the material goes below its freezing point. In contact with the burning material, the solid absorbs heat as it sublimes into a gas. In this way, carbon dioxide does provide some heat absorption, but it is minor in relation to its ability to smother the flames.

As with most extinguishing materials, carbon dioxide itself is not poisonous, but if its concentration in an enclosed area should go above 15 to 18 percent the oxygen content of the remaining air is insufficient for normal breathing. The result is partial asphyxiation: headache, gasping for breath, dizziness, sleepiness, and often a ringing in the ears. The best cure is to get out of the high concentration of carbon dioxide and back into a normal atmospheric mixture.

At one time carbon dioxide was used indirectly in fighting fires by being the propellant in both portable soda-acid extinguishers as well as the large chemical engines that saw service about the turn of the century. In these devices, the following reaction takes place:

$$2NaHCO_3 + H_2SO_4 \longrightarrow 2H_2O + Na_2SO_4 + 2CO_2 \uparrow$$

Interpreting this chemical shorthand, we find that two units of sodium bicarbonate (common baking soda) will react with one unit of sulfuric acid to produce two units of water, one unit of sodium sulfate (a salt), and two units of carbon dioxide (a gas).

In the portable fire extinguisher, a bottle of sulfuric acid is suspended in a wire basket over a solution of baking soda and water. When the unit is inverted, the sulfuric acid spills out of its bottle and mixes with the soda solution. The chemical reaction occurs and the gas is generated. The pressure that results from the generated gas forces the solution out through the hose and nozzle. As with all chemical reactions between inorganic acids (in this case sulfuric) and inorganic bases (sodium bicarbonate here) a salt and water are formed. The salt in this case is still in solution and so it and the water are forced out of the unit as the carbon dioxide pressure rises. Because the salt is in solution, it is ionized, and the whole solution is an excellent conductor of electricity that must not be used on fires where electricity in any form is involved.

A friend of one of the authors had an experience with just such an extinguisher in a college dormitory many years ago. A studio couch with a floor lamp at one end was involved in a fire. The floor lamp was plugged into an outlet directly behind the couch. Whether the electric cord was the cause of the fire or the couch was ignited by some other means is unknown but the cord became charred for almost its entire length. However, one wire continued to be insulated from the other until the stream from a soda-acid extinguisher hit it. Approximately three feet of the wire lit up as a brilliant blue arc for one or two seconds until the fuse blew and shut off the electricity. Fortunately the floor was dry and the extinguisher operator well insulated, or he too may have become a path for the electricity.

Before the discovery of electricity, the king-size soda-acid fire extinguisher served well for many, many years as the forerunner of the booster line for quick application to small and incipient fires, first as 20 or 30 gallon two-wheeled rigs, hauled by hand to the scene of the fire, then with larger horse drawn units, and finally in the form of one or two 35 to 100 gallon tanks carried on a motor-driven engine. The chemical unit did yeoman service until pumping from a tank of plain water became standard. Manufacture of even the small portable units was discontinued in the United States in 1969. Their place in the fire arsenal was taken by stored pressure and hand-pumped water types in some instances and by dry chemical and dry powder types in others.

DRY CHEMICALS AND DRY POWDER

While the terms dry chemical and dry powder are often used interchangeably, there is an official distinction made between the two.

The term **dry powder** is officially reserved for simple compounds used on fires involving solids of the class that cannot be fought with water, such as alkali metals. **Dry chemical** is correctly used for extinguisher compounds that are more complex in nature and usually used on burning liquid fires.

The most common of the dry extinguishing substances is sodium bicarbonate. When dry and under the influence of heat, such as from a fire, it reacts as follows:

$$2NaHCO_3 \longrightarrow Na_2CO_3 + H_2O + CO_2$$

Thus the reaction provides the carbon dioxide, some water, and a salt which acts as a heat sink. In this way energy is removed very rapidly, and further atmospheric oxygen is prevented from getting to the burning liquid.

Potassium bicarbonate is claimed to be even more effective as a dry chemical than sodium bicarbonate. If this is true the increased effectiveness has little or nothing to do with the increased chemical activity of potassium bicarbonate over sodium bicarbonate. However, it may well be due to the lower decomposition temperature, about 212 °F (100 °C), of potassium bicarbonate as compared to that for sodium bicarbonate at about 518 °F (270 °C). The lower decomposition temperature would cause the blanketing effect to occur much faster.

Several of the dry extinguishing mixtures have, as one of their important components, carbon in one of its crystalline forms known as graphite. The graphite crystals take the form of small platelets, which give the material a soft, black appearance, a greasy feel, an ability to conduct heat better than most nonmetals, and a tendency to adhere to solid surfaces. The high contact area of the platelets with each other produces a cooling effect on the fire by conducting heat away from the burning solids. This characteristic together with a blanketing effect, also attributable to the platelets, makes graphite quite effective as an extinguishing agent.

CARBON TETRACHLORIDE

Carbon tetrachloride is a noncombustible, and therefore unique, organic compound that has the chemical structure:

$$
\begin{array}{c}
\text{Cl} \\
| \\
\text{Cl} - \text{C} - \text{Cl} \\
| \\
\text{Cl}
\end{array}
$$

It was once thought to be an ideal fire extinguishing agent. We now know that when carbon tetrachloride is heated to decomposition, this halogenated hydrocarbon produces an extremely toxic gas, phosgene ($COCl_2$). While the specific heat (0.841 joules/g/°C) and the heat of vaporization (46.4 cal/g), as well as other characteristics, do very much recommend it for the quick quenching of small fires, the toxic effects demand that it be outlawed as an extinguishing agent. As an extinguishing agent, its low boiling point (about 170 °F or 77 °C) allows it to change from a liquid to a gas very quickly when it comes in contact with a fire. The heavy vapor spreads into the fire and shuts out the atmospheric oxygen. The problem is that phosgene forms just as quickly and in large quantities. Phosgene is so deadly that it is believed that 80 percent of the war casualties in World War I were due to its use by the combatants. It is made even more deadly by its pleasant odor, that of newmown hay. Concentrations above 50 parts per million can be deadly even after a very short exposure. There is not always a warning that a hazardous exposure has occurred. Phosgene seems to attack the lungs and there, reacting with the water product of the human internal oxidation reactions, becomes hydrochloric acid (HCl). This acid is directly injurious to the cells. A case has been reported where no effects were experienced for as long as 24 hours after the exposure. Then suddenly the exposed person reported an inability to breath and a burning in the throat and chest. The patient died about 36 hours after the exposure. It would seem that in some cases the danger from the fire could be less than that from the extinguishing agent. For this reason carbon tetrachloride extinguishers are illegal in most states.

FOAMS

While carbon dioxide is an excellent fire extinguishing agent, it does nevertheless have a distinct disadvantage—at normal temperatures it is a gas. As a gas it contains a tremendous amount of molecular activity and will take the shape of the vessel containing it. If that vessel is the whole world, the gas will wander off and no longer deny atmospheric oxygen to the combustion reaction.

To overcome this disadvantage, a product was developed that provides chemical vessels for the gas and makes it stay on its job in the fire. It was found that if a compound such as aluminum sulfate (Al_2SO_4) was mixed with sodium bicarbonate in a water solution, containing an agent which would increase the film-forming strength of the water, a foam would result. The foam is really a substance made up of millions of very small bubbles of carbon dioxide trapped by the strong films of water. The bubbles have such excellent cohesive strength that they stick together to provide a relatively stable blanket of carbon dioxide. The blanket, of course, prevents oxygen from getting to the fire.

In effect, foam provides two extinguishing media to the fire, carbon dioxide and water. The foam also overcomes one of the disadvantages of water noted earlier; it in effect does away with the normal low viscosity of water, which allows it to drain away too rapidly from a burning substance.

TYPES AND CHARACTERISTICS

Foams are essentially delicate mechanical structures based on ultrathin films of some liquid. Anyone who has ever blown soap bubbles by means of the old bubble pipe technique, or even the more modern bubble ring method, will attest to the delicacy of the film. For this reason, fire extinguishing foams will break down if not properly used.

There are many types of foams, each developed to fill a need in fire extinguishment. In addition to the simple acid-base type mentioned earlier, which has now become obsolete, there are now the more effective organic type foams in widespread use:

- surface-active foaming agents
- soluble protein foaming agents
- combinations of surface-active and amino acid foaming agents
- aqueous film-forming foaming agents
- alcohol foams
- high expansion foams

Surface-Active Agents

Surface-active agents are organic compounds that are soluble in water, and when dissolved in amounts up to about 5 percent they drastically reduce the surface tension of the water. This characteristic increases their wetting properties. They are sometimes called surfactants.

In the proper equipment, many of these surface-active agents will produce good volume foams but they tend to be weak and to break up very easily. Because they are organic agents they tend to be incompatible with other types of foams and therefore cannot be used together with them or from the same foam-making equipment, unless it is thoroughly cleaned between solutions.

The foam resulting from these organic surfactants has a very low viscosity and excellent covering characteristics on solid as well as liquid surfaces. Its value as an extinguishing agent lies in its ability to keep atmospheric oxygen from the fuel and so stop the combustion process. Because these surface-active agents have little or no film-forming properties they must be used in large volumes in order to achieve success.

Anyone who has ever washed clothes or dishes is familiar with these foams, and is aware of how fragile their structures are. Both soap and detergents are surface-active agents and, as such, reduce the surface tension of the water. This reduction in surface tension allows them to perform other tasks besides making foam and emulsifying dirt, as we shall see later in this chapter.

Because detergents are synthetic organic compounds, they can be tailored to many different tasks. At this time there are over 300 surface-active agents available in the United States. Most surfactants have chains 10 to 15 carbon atoms in length and are usually processed from coconut or palm oil. These synthetic detergents are resistant to hard water and more stable than soaps to solutions with low pH.

Soluble Protein Foaming Agents

This type of foaming agent is formed from long chains of molecules linked together in such a way as to form huge, spherical cross-linked polymers. While the molecular weight of most inorganic compounds is well below 100, the molecular weight of the foam polymers can be in the hundred thousands. One of the largest molecules ever seen by means of an electron microscope is the tobacco mosaic virus (a true protein) whose molecular weight has been measured as being between 15 and 20 million.

Without additives, soluble protein foams tend to be temperature sensitive, unstable, and less than optimum volume-producing agents. With additives, such as specific metallic salts and soluble organic foaming agents, they can be highly stable and generate large volumes from small, concentrated solutions. A favorite protein foam of many firefighters is the meringue on lemon meringue pie, an excellent example of a relatively stable protein foam. It is made of a soluble protein, egg albumin, together with sugar (for taste) and cream of tartar (an organic potassium compound) for stability. While this foam may not have put out any flame-emitting combustion reactions, it has surely stoked plenty of internal oxidation-reduction reactions.

Other well known soluble proteins are casein, gelatin, serum globulin, and insulin.

Combinations of Surfactants and Protein Types

The basic soluble protein foams have a disadvantage that manifests itself in a liquid fuel fire. The flammable hydrocarbon tends to coat the protein foam bubble and so destroys it. However, certain fluorine derivative surface-active agents, when mixed with the protein foams, will produce a foam of improved stability and excellent cohesive characteristics that tends to contain the flammable vapors in an improved manner.

Aqueous Film-Forming Foam Agents

Aqueous film-forming foaming agents are surface-active fluorinated carboxylic acids with additives to improve their film strength and stability.

Unlike many chemical names, carboxylic acid is not the name of a single compound but rather the name of a whole family of compounds. They all possess the carbonyl group ($C=O$) and a hydroxyl group (OH). As with so many organic acids, they have only weak acidic characteristics. In inorganic systems, that hydroxyl group would be the signal of an alkaline compound; but this is not so in organic compounds. The contrast comes about because inorganic compounds are almost completely ionized (more than 95 percent) while organic compounds are almost completely nonionized (less than 0.5 percent). There are some hydrogen ions present in the organic acid and so it is only a very weak acid. One of the most common members of the carboxylic acid family is vinegar. Its chemical name, when it is pure, is acetic acid. The formula for acetic acid is

$$
\begin{array}{ccc}
\text{H} & \text{O} & \\
| & \| & \\
\text{H} - \text{C} - \text{C} - \text{O} - \text{H} \\
| & \\
\text{H} &
\end{array}
$$

If acetic acid were to be fluorinated, as is the carboxylic acid used in aqueous film-forming foam agents, it would become fluoroacetic acid and have the formula

$$
\begin{array}{ccc}
\text{H} & \text{O} & \\
| & \| & \\
\text{F} - \text{C} - \text{C} - \text{O} - \text{H} \\
| & \\
\text{H} &
\end{array}
$$

Foams produced by the fluorinated carboxylic acid and additive combinations have excellent spreading and barrier characteristics, even in face of heavy fuel vaporization. These agents foam with simple apparatus and are quite compatible with other types of extinguishing foams. To avoid the tongue twister "aqueous film-forming foaming agent," they are often called, "A3F," "A triple F," and even "A-F-F-F" agents.

Alcohol Foaming Agents

Because air bubble containing foams are relatively unstable and therefore ineffective in the presence of water soluble hydrocarbons such as alcohol, acetone, ether, and lacquer thinners, specific foaming agents have been developed that are really a subclass of the soluble protein/surface-active foaming agents discussed earlier. These agents are *not* alcohol-based foaming agents, but are for alcohol fires.

There are three general classes of these agents:

- Multiple component systems consisting of a partial polymer and a catalyst, which will promote the complete polymerization some time after the components are mixed.
- Single component systems, which can be used on any flammable liquid as well as water-soluble hydrocarbons.
- Soluble protein-alkali soap systems made soluble in water by means of special additives. This type has distinct disadvantages: because of its reaction time, which is relatively short, its use is limited to the length of line it can be pumped through; and it must be applied gently to the flaming liquid to be at all effective.

High Expansion Foam Systems

These are really specialized types of surface-active foaming agent systems that will provide expansion ratios of up to 1000 to 1. They are especially effective in control of enclosed fires, where the space is limited to a small volume. With this type of agent, the foam can be used literally to squeeze the confined smoke and vapors out of the area and so engulf the complete space whether it is easily accessible or not to an extinguishing agent.

EQUIPMENT

Because foam is really a delicate mechano-chemical structure, it does not exist, as does water, in nature. It must be manufactured. Not only are the chemical compounds discussed previously required to make it, but these compounds must be used in specialized equipment that will produce the desired structure. These systems may be classified as mechanical or chemical.

The mechanical systems may be further subdivided into direct and indirect pressure systems.

Direct Pressure Systems
Air Type

In these devices air is pumped directly into the foam concentrate. Under sufficient pressure stable foams can be developed and pumped onto the flames.

Water Type

These devices are usually simple spray nozzles into which water, containing special additives, is pumped. They depend upon nozzle turbulence and stream breakup to form the foam. While they are simple, they are not very satisfactory for widespread applications. Many of the aqueous film-forming foams are used in this system.

High Expansion

There is one type of high expansion foam system that uses direct air pressure and direct foam pressure to deliver large amounts of foam to enclosed areas. In this device the foam must be preformed before it is injected into the air stream. A schematic diagram of this system is shown in Figure 11.1.

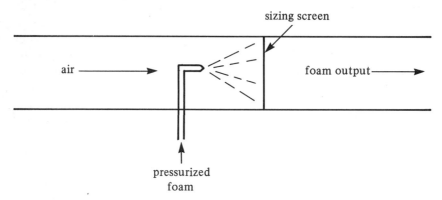

Figure 11.1. High Expansion Foam System.

Indirect Pressure Systems

Atmospheric

Atmospheric systems make use of the Venturi principle to pump concentrate out of its container and mix it with a stream of water in some specific ratio. The Venturi principle is basically that which is used on the end-of-the-hose garden sprayers to deliver insecticide for home owners. In this system the stream of water is passed over an orifice at high velocity. The passage of the high velocity water stream produces a lower than normal pressure at the orifice. A pipe is connected from the orifice to the concentrate in a container that is open to atmospheric pressure. The lower pressure at the orifice causes the concentrate to be pushed up from its container and out of the orifice, where it mixes with the water. The pushing is caused by atmospheric pressure acting on the surface of the concentrate and forcing it into the lower pressure area at the orifice. An atmospheric system is presented schematically in Figure 11.2.

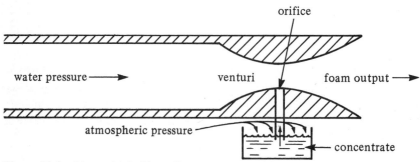

Figure 11.2. Atmospheric Foam System.

While the actual construction of these aspirating mixers will vary not only in design but also in complexity, the basic principle remains the same: a high velocity stream of water develops a reduced pressure area causing mixing of the concentrate forced into the stream by atmospheric pressure.

This type is usually referred to as an eductor, from the Latin word *educere,* meaning "to lead out."

Boosted Pressure Type

When the stream velocity is reduced, either because of long lines or because the foam must be injected beneath the surface of a burning liquid, the simple aspirating system must be replaced by a high back-pressure system. In these devices the lack of pressure differential is overcome by putting pressure on the tank of concentrate. These devices are more complex but will supply a continuous stream of foam as long as the pressurized reservoir contains concentrate.

High Expansion Aspirating

In the simple Venturi foam generating system, water flowing by an orifice developed the low pressure area that caused the foam concentrate to be pushed out of its tank. There are some high expansion systems that use the Venturi principle but with air providing the velocity to produce the reduced pressure. These devices are especially efficient and produce a very high foam to concentrate ratios (up to 1000 to 1) in a special turbulence chamber before passing through a bubble sizing screen and out of the nozzle. This type is shown schematically in Figure 11.3.

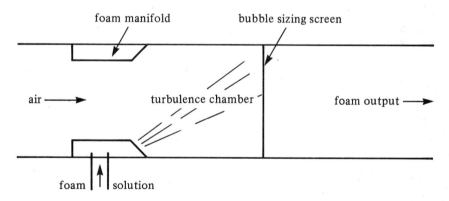

Figure 11.3. High Expansion Aspirating Foam System.

Chemical Systems

In the chemical systems, chemical reactions rather than mechanical devices produce the foam. For this reason the specific recipe for a given mixture must be carefully followed in order to produce optimum results. These are for the most part three component systems: two solids and water. The two solids are usually in powder form to provide large surface area for reaction speed. Gravity feed is usually used for the powders but a Venturi throat is used to increase the velocity of the water. This improves the mixing action and speeds up foam production. The Venturi throat is merely a curved narrowing and more gradual widening of the internal diameter of the line.

The powder may be premixed in a single hopper or may be fed from dual hoppers. It would seem to be a simple operation to let powder fall out of a hopper into a stream of water but the actual systems can be very complex, because they are designed to provide optimum mixing and yet prevent gumming of the powders at their outlet or outlets into the jet of water. When gumming occurs the device must be completely disassembled, cleaned, and thoroughly dried before it can be used again.

HALOGENATED EXTINGUISHING AGENTS

Earlier in this chapter we discussed carbon tetrachloride as an outdated and, in many states, illegal extinguishing agent. You will recall that it is illegal, not because it did not extinguish fires, but because of the toxicity of the gas it produced when it began to decompose in the heat of the fire.

While foams do an excellent job of blanketing and extinguishing a fire, they do have at least one very great disadvantage: they create a tremendous mess. Imagine, if you will, the mess a computer center fire would leave if it were extinguished with either water or foam! If a compound could be developed that was a gas (and so leave no mess), would have good self-adhesion (cohesion) and so hang together like a blanket and deny oxygen to a fire, would not decompose leaving toxic compounds—that would be an ideal extinguishing agent for many fires that cannot take water.

In the late 1950s the Army Research and Development Command at Fort Belvoir, Virginia, developed a number of compounds related to carbon tetrachloride but without its highly toxic products of decomposition.

Some of the more promising of these are:

Name	Formula	Structure
bromochloromethane	CH_2BrCl	$\begin{array}{c} Br \\ \vert \\ H-C-H \\ \vert \\ Cl \end{array}$
dibromodifluoromethane	CBr_2F_2	$\begin{array}{c} Br \\ \vert \\ F-C-F \\ \vert \\ Br \end{array}$

bromochlorodifluoromethane	$CBrClF_2$	Br | F — C — F | Cl
bromotrifluoromethane	$CBrF_3$	F | F — C — F | Br
dibromotetrafluoroethane	$C_2Br_2F_4$	Br Br | | F — C — C — F | | F F

Not expecting everyone to remember those long chemical names, the Army Corps of Engineers called the compounds simple **halons,** from their chemical family, halogens. In order to distinguish one from the other they devised a simplified numerical system, under which the extinguishing agent is given a four or five digit number and is called halon XXXX.

The Corps of Engineers did not empirically pick numbers from a random number chart. Each of the halon numbers is coded to the chemical compound. The first digit in the number indicates the number of carbon atoms in the molecule, the second digit the number of fluorine atoms, the third digit the number of chlorine atoms, the fourth digit the number of bromine atoms. A fifth digit, when needed, indicates the number of iodine atoms.

For example:

Chemical Name	*Halon System Name*
bromochloromethane	halon 1011
dibromodifluoromethane	halon 1202
bromochlorodifluoromethane	halon 1211
bromotrifluoromethane	halon 1301
dibromotetrafluoroethane	halon 2402

This order, established by the Corps of Engineers, is quite logical. Carbon is numbered first because it is the basis of all organic molecules. The halogens are then listed in the order of their atomic number. Fluorine has an atomic number of 9, chlorine 17, bromine 35, and iodine 53. Another way to remember the halon order is the mnemonic, *C*alling *F*irefighters *C*ancels *B*lazes *I*nstantly, where C is for carbon, F for fluorine, C for chlorine, B for bromine, and I for iodine.

A chemist might be upset by the lack of a number for hydrogen atoms in these organic compounds. The Corps of Engineers decided, however, that since any carbon linkages not satisfied with halogen atoms would naturally be filled with hydrogen atoms, just as they are in all saturated organic molecules, it was unnecessary to note the number of hydrogen atoms. They also decided that any zeros that would appear at the end of the number, indicating a lack of iodine in the molecule, would also be dropped.

Although the popularity of halons on the fire scene is fairly recent, they have been around for a long time. The first, carbon tetrachloride, now sometimes referred to as halon 104, appeared in Europe just before the turn of the century.

Allied aircraft in World War II used methyl bromide ($CH_3 Br$), now called halon 1001, for engine fire protection. Again, as with carbon tetrachloride, the decomposition products are highly toxic. While this was no problem with in-flight engine fires, where there was plenty of ventilation, it could not be accepted for enclosed area fire-fighting.

Axis aircraft of World War II used a halogenated compound called CBM for chlorobromomethane ($CHClBr$). While this compound is far superior to methyl-bromide, it still develops highly toxic decomposition products under flame conditions. Therefore this halon 1011 could not be accepted either for enclosed fire-fighting. The obvious potential of these halogenated hydrocarbons excited the imagination of research scientists both within the United States Government and private industry in the United States as well as many European countries. They set about to at least reduce, if not eliminate, the toxic decomposition products.

By 1950 most of Europe had adopted halon 1211 because it was far less hazardous and far more effective than the parent halons, 104 and 1011. Another halon, 1202, is finding favor for aircraft systems because it carries considerable fire protection in a small package. It has a high liquid density and excellent blanketing characteristics.

In the United States, halon 1301 has a wide edge as the favorite, while another halon, 2402, is gaining rapidly in popularity. In fact, claims have been made for halon 2402 that someday it or one of its ethane relatives will replace many of the foaming agents for liquid fire extinguishing.

HALON EXTINGUISHING—
THE MECHANISM

There have been several hypotheses advanced, some even dignified with the title of theory, as to how the halons extinguish fires so well. There have been suggestions of complex free-radical development and of ions, formed in the combustion process, being inhibited by the halons.

A free radical is the reaction chemist's way of explaining a broken covalent bond or linkage that results in an unpaired electron on a combination of atoms. This has been found to be necessary if a reaction mechanism is to be developed for some of the more complex chemical reactions. For example, R—O—O—R is said to form the free radicals R—O · and · O—R, which then can react very quickly with other molecules. Free radicals are extremely reactive, and never exist under nonreaction conditions.

Free radicals most certainly exist, but whether this is the mechanism of halon extinguishment has not been proven at this time. All of the proposals concerning halon extinguishment are really just hypotheses in the scientific sense; very little experimental work has been published that confirms or denies such conclusions.

The basic problem in any extinguishing mechanism is the combustion mechanism itself. As noted in Chapter 2, solids and liquids, as such, do not react rapidly enough to be thought of as taking part in a combustion reaction. Until considerably more objective evidence is published, it would seem unwise to strain for complex explanations—it could be the phlogiston theory all over again. The simple, and admittedly even oversimplified, view that initiation energy transforms enough of the solid or liquid into gas molecules, which then react, explains a considerable number of observations. The objection that wood does not form a liquid before becoming a gas before burning can be explained by the view that wood, at or below its ignition temperature, undergoes a form of sublimation similar to carbon dioxide at room temperature. It is not necessary that sublimation take place only at low temperatures. The form of sublimation for wood might be called **pyrolitic sublimation**.

The further objection that extinguishing techniques, such as explosive pressure waves, resonant vibration, or strong electric fields, cannot be explained in light of simple oxidation of gas molecules is unfounded. All of these phenomena have been shown to effect gas reactions even to the point of removing gas molecules, thereby removing excessive exothermic energy. The result is a general cooling down of the fire system and prevention of further gas development.

The simple explanation, in light of the lack of objective proof or more complex mechanisms, is in no way meant to detract from the halons: they work! Consider their excellent characteristics. As the number of halogen atoms on the molecule increases, the halons can have almost twice the density of carbon dioxide and about half again as much as water. The density of their vapors approaches five times that of air, and all halons have a low heat of vaporization, a low boiling point so that they are able to produce a heavy, blanketing gas at relatively low temperature. They are liquids with freezing points well below −200 °F (−129 °C), so they can be used at high altitude for aircraft or in very cold weather without fear of freezing.

TYPES OF FIRES

It should be obvious by now that there is no one extinguishing agent for all fires. To make it easier to chose the proper type of agent for any particular fire, the NFPA Extinguisher Standards classify fires as follows:

Class A: Fires involving ordinary combustibles: wood, paper, cloth, and some plastics.

REQUIRED: Cooling from water or water solutions.

Class B: Fires involving flammable liquids, flammable gases, grease, etc.

REQUIRED: Oxidizer exclusion.

Class C: Fires involving live electrical equipment. When electrical circuits to the equipment are not energized, the fire may be treated as a class A or class B fire. It is not enough that the equipment merely be switched off. The electrical feed circuits to the equipment must be off or disconnected.

REQUIRED: Nonconductive cooling and/or oxidizer exclusion.

Class D: Fires involving combustible metals of the alkali family: magnesium, potassium, sodium, titanium, zirconium, etc.

REQUIRED: A nonreactive, energy-absorbing agent.

REVIEW QUESTIONS

1. Give three advantages and three disadvantages of water when used as an extinguishing agent.
2. Define specific heat, heat of vaporization, and surface tension.
3. What is wet water?
4. What is slippery water? What mechanism has been proposed to explain this phenomenon? What advantages does it produce?
5. Explain in detail some of the reactions that water promotes.
6. What is carbon dioxide? How can it be used as an extinguishing agent?
7. Define dry chemical and dry powder extinguishing agents. What is the distinction between them?
8. Why should carbon tetrachloride *not* be used as an extinguishing agent?
9. What is a foam? Name three of the more popular types of foam extinguishing agents.
10. What is an alcohol type foaming agent?
11. Name the types of direct pressure foaming systems.
12. Name the types of indirect pressure foaming systems.
13. Explain the Venturi principle.
14. Explain the derivation of the term halon.
15. Explain the halon numbering system.
16. What is a class A, class B, class C, and class D fire? What is required for extinguishment of each?

Electricity/ 12

Any hazard that does not activate one of the senses of man with some sort of alarm must be considered a greater hazard because it gives no warning. More often than not its hazard potential will be feared more than it should be just because it is unseen. While firefighters, as a group, have never been known to run away in fear, they do know that they must learn to respect certain hazards whether these hazards give warnings or not.

Nuclear radiation is one hazard that does not warn. However, an even more common one is electricity. Harmful exposure to nuclear radiation is relatively rare, but almost every place is electrified and as such, contains the serious, deadly hazard of electric current.

There are two common types of electricity that can perform work and so be a serious hazard: alternating current and direct current. Originally direct current was the only type used because it was the simplest to generate. As the science progressed, it was found that direct current could not be transmitted over great distances as could alternating current. Modern systems use alternating current. Although some older buildings in older parts of larger cities still have direct current in them, they are quickly becoming rare.

Whether it is direct or alternating current, electricity may be defined as a flow of electrons from a source, or generator, through a load (such as a motor or lamp) and back to the source. In order to make the electrons flow through that path it is necessary to have a supply of electrons, enough pressure behind that supply, and conductors that will allow the electrons to flow from the source to the load and back again.

CHARACTERISTICS

SUPPLY AND CURRENT

The number of electrons that a generator is capable of producing determines, with the load, how much electron flow will exist in a conductor. This quantity of electrons is called the **current** or **amperage**. It is this characteristic of electricity that does the damage to the human body.

It must be understood that without the pressure behind it there can be no current.

PRESSURE

The amount of pressure behind the flow of electrons is called the **electromotive force** or, more commonly, **voltage**. This flow is directly related to the ability of the generator, whether it be a mechanical type or a chemical type, such as a battery. Voltage is similar to head pressure in hydraulics. When the water supply is high above the nozzle, the pressure is great. This is similar to high voltage. When there is very little differential in height between the supply and the nozzle there is very little pressure. This is similar to low voltage. Many believe that high voltage kills. It does not. It is the amount of flow, or current, that does the damage.

CONDUCTORS

The atoms of some elements or substances will allow one or more of their outer orbital electrons to be displaced with relative ease, while other materials will not. Those that resist the displacement of their electrons are called **insulators**. Materials that allow their electrons to be knocked off with ease are known as **conductors**. There are all degrees of conductors and insulators depending upon how much resistance the outer shell electrons offer to the displacement mechanism. In general, metals and inorganic acids are good conductors, while organic materials are usually nonconductors.

The resistance that any given material offers to the flow of current is dependent upon: the *material* (element or alloy); its *cross-sectional area;* the *length* through which the current must flow to return to the generator; and the *temperature* of the conductor.

MATERIALS

In order of increasing resistance, the more common conductors are:

Material	Relative Resistance
silver	1.00
copper	1.05
gold	1.50
aluminum	1.74
tungsten	3.39
zinc	3.53
nickel	4.26
iron (electrolytic)	6.11
platinum	6.73
steel (transformer)	6.82

The difficulty a free electron has in displacing an outer shell electron of an adjacent atom is directly related to the resistance. The energy that is lost in the displacement collision appears as heat, causing a rise in the temperature of the conductor—as long as the current demand by the load is there. In order to reduce this resistance and therefore the heat rise in the conductor, it would be ideal if all conductors were made of silver. However, the cost of this element is so great that it becomes impractical for most uses. The abundance of the element copper, along with its relatively low resistance, makes it the best compromise. Actually it is not much of a compromise when we realize the copper is only about 5 percent poorer as a conductor than is silver.

CROSS-SECTIONAL AREA

Because greater cross-sectional area means more atoms, and therefore more electrons, to help conduct the current, the resistance of a conductor increases as the wire gets smaller in diameter. Conversely, the resistance becomes less as the diameter is increased.

Most firefighters at one time or another have been on jobs where the cause of the fire was an extension cord which became so hot it ignited adjacent combustible materials. This is an excellent example of a cross-sectional area that is too small for that current flow offering too much resistance to that flow, with the resultant loss of energy appearing as heat.

LENGTH

As with cross-sectional area, the amount of work the current must do to get to the load and back to the generator causes a loss in energy. If the flow must be over great distances, the resistance caused by the long lines produces considerable loss in voltage if the current flow is high. This situation is aggravated when small wires are used over long distances. Then the length and the cross-sectional area both result in energy losses and a significant rise in conductor temperature.

If the current for long distance electrical transmission is kept low, the loss is also low. For this reason cross-country transmission lines or any high wattage distribution system operates at very high voltage so that the currents stay low for any given wattage consumption. For example, a 1000 watt flood lamp at 12 volts will demand about 83 amperes. A 1000 watt lamp at 440 volts would demand only about 2 amperes. Imagine wattage demands in the million watt region, and it is not hard to understand why long distribution systems run with tens of thousands of volts.

TEMPERATURE

As we learned earlier, the activity of the atoms increases as their temperature increases. This increased activity results in a greater resistance to outer shell electron displacement and so a greater resistance to current flow. The higher the temperature of the conductor the greater the electrical resistance.

This effect is quite evident in an extension with conductors that are too small for the power demand. As the conductor temperature increases, the resistance of these conductors increases causing even greater energy loss and a resultant temperature increase. The cycle continues to destruction of the material by melting, or at least until the flow is stopped.

The amount of resistance increase per degree rise in temperature is dependent upon the specific material from which the conductors are formed.

There are also a few materials that decrease in resistance as the temperature increases, but they are relatively rare and expensive. For this reason they are used only in specialized applications and need not concern us.

DIRECT CURRENT

With the simple analogy of pressure and water flow, it is only natural that we think of the flow of electrons in one direction from the generator through the load and back to the generator. Such electricity does exist and is known as direct current, more often than not abbreviated as DC. This type of current is the only type of electricity generated by chemical methods: batteries. DC generators of the mechanical type also exist, and in fact were once the only type used in automobiles and trucks to recharge the batteries.

ALTERNATING CURRENT

It is also possible to have an electromotive force or pressure that periodically reverses itself so that the current first flows in one direction and then completely reverses itself and flows in the opposite direction. This is called alternating current, and is usually abbreviated AC. The number of times the pressure reverses itself in one second is called the frequency. This frequency is designated in cycles per second or cps. In modern terminology it is now labeled as hertz, or Hz. The old 60 cps then becomes 60 Hz. Older equipment may be marked as cps, but many years ago it was decided to distinguish between cycles per any unit of time and cycles per second by combining the cycles per second into one label, hertz, the name of a German scientist famous for his early work with alternating currents.

OHM'S LAW

Whether the current flow is direct or alternating, the electric circuit and its more common characteristics can be expressed as a relationship between the voltage, the current, and the resistance of the circuit to the flow of electrons. This relationship is known as **Ohm's law** and may be expressed as follows: the electron flow (current) in a circuit is equal to the voltage drop across the circuit divided by the resistance of the circuit. Mathematically it is expressed:

$$\text{Amperage} = \frac{\text{Voltage}}{\text{Resistance}}$$

More often than not you will see it written $I = E/R$, where I is the *in*tensity or current, E is the *e*lectromotive force or voltage, and R is the *r*esistance.

Because it is a mathematical expression it may also be solved for E or R: $E = I \times R$, or $R = E/I$.

example: If a lamp demands 8.5 amperes of current when across a 12 volt line, what is its resistance?

$$R = \frac{E}{I} = \frac{12}{8.5} = 1.4 \text{ ohms}$$

example: At what voltage will a 10 ohm resistance demand 15 amperes?

$$E = I \times R$$
$$E = 15 \times 10$$
$$E = 150 \text{ volts}$$

While Ohm's law applies to both alternating and direct current, it becomes far more complicated for alternating currents where the reversal of flow produces effects that must be considered for complete accuracy. Nevertheless, for our work, Ohm's law can be considered sufficiently accurate for both types of current.

There is another very important characteristic of current flow in a circuit. This is the rate at which the electrons do the work, known as **power**. The unit of power is the **watt.** Because the two components of flow, voltage and current, determine how much work is performed, a watt is equal to the voltage multiplied by the current when the voltage is in volts and current in amperes. This power formula is expressed as $P = E \times I$, where P = watts, E = voltage, and I = amperes.

example: How much current flows in a circuit lighting a 60 watt lamp across a 120 volt line?

Put the given values in the equation:

$$P = E \times I$$
$$60 = 120 \times I$$

Solve the equation for I:

$$120I = 60$$
$$I = \frac{60}{120}$$
$$I = 0.5 \text{ amperes}$$

However, that same 120 volt line with a 1000 watt lamp across it would have:

$$P = E \times I$$
$$1000 = 120 \times I$$
$$120I = 1000$$
$$I = \frac{1000}{120}$$
$$I = 8.3 \text{ amperes}$$

The important point to be gained by these examples is that although the voltage may be the same, the current varies with the load. The only limits to the amount of current flow are the ability of the generator or voltage supply; the current carrying capacity and melting point of the conductors; and the ratings of the protection devices in the circuit.

PROTECTION DEVICES

FUSES AND CIRCUIT BREAKERS

Both of these devices perform the same service: to protect the generator and supply conductors from overload.

The original protective devices, fuses, are designed about the simple observation that the cross-sectional area and the material of a wire determines how much current it can carry before the temperature rise of the wire reaches the melting point of the material. When the melting point is reached, the conductor melts and opens the circuit like a switch. Of course, it is a one-shot switch and must be replaced with a new fuse after each opening.

By choosing the proper metal alloy for its melting point, along with the proper cross-sectional area, a fuse can be designed to melt and open anywhere from a few hundredths of an ampere to hundreds of amperes.

The newer protective devices, circuit breakers, can take several different forms but there are two basic types, the bimetal and the magnetic.

The lower current devices work on the principle that two metals, having sufficiently different heat expansion ratios and placed back-to-back, can be made to snap like a switch and open the circuit when the current rises above the ratings and the resultant heat causes the two metals to expand differently. They can be designed for various current flow operations, and have an advantage over fuses in that circuit breakers need not be replaced after each overload. A simple resetting, once the excessive current demand has been eliminated, is all that is needed.

The higher current devices, such as those used primarily in heavy industrial applications, are usually of the electromagnetic type. These devices work on the principle that the amount of electromagnetism that will develop is proportional to the current that flows for any given voltage. By proper design the magnet can be made to operate a switch mechanism and open the circuit when the current flow exceeds the designed value. These also may be reset once the excessive current problem is solved.

Because these heavy duty circuit breakers are handling such high current flow, some precaution must be taken to protect against excessive arcing as the circuit is opened. Heavy arcing destroys the mechanism in very short order. Some of these magnetic devices are immersed in special oil baths. The oil does not ionize as does the air and so no arc is developed. There are others that use a blast of air or some other gas to blow out the arc as it trys to develop.

HAZARD POTENTIAL

With any potentially hazardous material, it is the situations that are caused by the material or its contacts with the human body that remove it from the "potential" class and places it in the "now" hazard class. Electricity is no different.

While it is not always possible to evaluate, with accuracy, the potential electrical hazard that may exist in an area during a fire, it is possible to estimate whether the hazard is high voltage or high current, or both. This estimate can be made by reading certain indicators. These indicators are the types of conductors and insulators needed for high current or high voltage systems.

HIGH VOLTAGE

Just as high pressure in a fire hose will burst out anywhere a weakness in the containment exists, so the high pressure of an electrical system will burst out in the form of an arc or spark wherever an insulation weakness or just too much pressure exists.

There are many materials that are normally considered good insulators at normal voltages, yet will break down and leak under high voltage or high frequency conditions. Even the best materials need distance to provide adequate separation of a high voltage conductor from a conductor of zero level or of considerably lower voltage level.

Some of the more common insulators are: dry air, dry wood, porcelain, glass, rubber, and plastics.

Ribbed, conical insulators of porcelain, glass, or plastic indicate high voltage. The bigger the insulator, the higher the voltage the conductors may be carrying.

If at all possible, metal ladders should be kept away from such conducting systems. Even dry air will break down if the voltage is high enough and the distance to a zero voltage conductor small enough. A metal ladder at ground potential even on dry concrete is an invitation to disaster if it comes close enough to a high voltage conductor.

An indication of how close "close enough" really is may be seen in the following table for dry air and pointed electrodes. These conditions simulate what can be expected with metal ladders and high voltage conductors. Any moisture in the air increases the distance that a given voltage can jump.

Voltage	Maximum Dry-Air Breakdown Distance	
	in.	cm.
5,000	0.17	4.3
10,000	0.33	8.4
15,000	0.50	12.7
20,000	0.70	17.8
25,000	0.87	22.1
30,000	1.06	26.9
35,000	1.26	32.0
40,000	1.77	44.9
45,000	2.00	50.8
80,000	4.37	111.0
100,000	6.10	154.9
150,000	10.27	260.8
200,000	14.10	358.1
300,000	21.53	546.8

If you must go near high voltage systems such as the above, *three times* these distances is not too safe a minimum.

Too many fire safety personnel have been killed when a metal ladder they were putting up or taking down merely came within the breakdown distance of a high voltage line—a line they may not have suspected was there—and the ladder drew a spark. The current flowed through their bodies because they were in contact with the ladder. The high current stopped the heart muscle, and they died.

Systems

While power generating stations may generate the voltage at about 22,000 volts, it is usually increased to as high as 345,000 volts by means of transformers. A system has been built with pressures as high as 750,000 volts.

The overland transmission voltages are lowered in substations to any usable voltage, but it is not uncommon to have voltages in the order of 34,000 volts in such stations. Often these stations are associated with an industrial complex, but not exclusively so.

Industrial areas may receive voltages of about 14,000 volts and higher and drop it to 2000, 440, 220, or 120, as needed. Most homes and stores will have voltages at 220 or 120. This level is none the less deadly when direct contact is made.

HIGH CURRENT

The current in an electrical circuit is, like nuclear radiation, a silent killer. While it does not radiate, so that a firefighter must come in contact with it, electricity is so common that it is far more usual for people to be killed by current than by radiation.

Because of the relationships described by Ohm's law, the indicator of high current is the size of the conductors; a larger conductor has a higher current carrying capacity. This does not mean that small wires are not deadly! Common operating voltages even as low as 6 volts can kill if there is sufficient current available. *Any* energized electrical circuit should be considered an extremely hazardous situation and treated accordingly.

High current conductors may or may not look like wires. In industrial situations conductors may look like copper or aluminum bars of just about any dimension, called **buses**, running along walls or ceilings. Normally they will be in protective channels, but the channel is not always protective enough and contact can be made with deadly results.

Unlike high voltage, which may jump out of its conductor and do its damage, high current will stay in its conductor until direct contact is made. In addition, the high current will flow through the path of least resistance, which can be a metal ladder—but don't risk your life on it. The voltage may divide, with a sufficient amount taking the path through the body. While the resistance of the ladder will be close to an ohm or so, the resistance of the human body, which varies with individuals, is in the 50,000 ohm range when measured from hand to hand.

Doctors tell us that this path for electricity, from hand to hand, is the deadly one because the current flow, following the blood stream as the path of least resistance, is through the heart. The massive flow of electrons overloads this muscle and destroys it. For this reason, the prudent person who must contact a possibly live electrical circuit does it with one hand behind his or her back or in a pocket. An even better rule is: *if it might be hot, don't touch it.*

A special technique has been developed using electrical energy to restore normal heart rhythm once it has ceased or become seriously irregular. This technique does not use the hand-to-hand path, and uses high voltage but with low current as applied from a charged electrical reservoir called a capacitor. The method is very successful when used by an expert.

WATER AND ELECTRICITY

Water, the firefighter's friend, can become the firefighter's enemy in voltage and current to the unwary person who becomes part of the circuit.

The authors know of one case of a boat owner who, wanting to mount a name plate on the transom of his craft, decided to stand in the water to do it. With his bathing suit on and his metal-cased electric drill plugged into the dock, he jumped into the water, which came up to about his waist. Setting the drill point against the back of the boat, he pulled the trigger switch, and died. His drill had a short to the case in it and that put his body into a 120 volt, 100 ampere circuit: service wires to metal case, to his hands, to his body, to the water, to the ground.

INDUSTRIAL ELECTRICAL HAZARDS

Industrial electrical hazards are, for the most part, high current systems with voltages ranging from 6 to 440 volts. They use both alternating and direct current. Input services to the industrial area may run in the region of many thousands of volts with very high current.

A few of the more common applications in industrial situations are:

Application	Voltage	Type	Current
plating	low	DC	high
arc furnace	high	DC	high
welding	low	DC	high
drive motors	medium	AC and DC	varies with horsepower
electric trucks	low	DC	high
battery chargers	low output	DC	high
transformers	variable	AC	variable
electric signs	high	AC	variable
electronic equipment			
solid state	low	DC	variable
tube type	high	AC and DC	medium

REVIEW QUESTIONS

1. What are the two working types of electricity? Discuss their differences.
2. What is the current, or amperage? What does it describe?
3. What term is used to describe the electron pressure? What is EMF?
4. How does a conductor differ from an insulator?
5. What is resistance in an electrical circuit? What is the unit of measurement?
6. What characteristics of a wire influence its resistance, and in what manner?
7. Give Ohm's law in both plain language and mathematical terms.
8. If a lamp draws 12 amperes with 120 volts across it, what is the resistance of that lamp?
9. In a 12-volt circuit, 10 amperes flow when a switch is closed. What is the load resistance? What would the current be if only 6 volts were applied to the circuit?
10. What would the current be in a 12-volt circuit where the load has a resistance of 5 ohms?
11. What is the relationship between power, voltage, and current? Express it in mathematical terms.
12. Explain in detail the differences between fuses and circuit breakers.
13. What are the telltale signs of high voltage, high current, low voltage, and low current?
14. What is the deadliest path for electricity in the body? What precautions should always be taken near exposed electrical conductors?
15. How does water increase the hazards of electricity?

Radiation Hazards/ 13

Fire, a visible form of radiation, is energy developed by the orbital components of atoms. When under control, it can serve us well,·but out of control it can become a hazard to life and property. Nuclear radiation is also a form of energy but one that is developed by internal atomic components. As with other forms of energy, nuclear energy can serve us well, but when out of control, it, too, becomes a hazard to life and property. Unfortunately, we do not always know when it is out of control, because the human body has no sensors (eyes and nerve endings) for nuclear radiation as it does for some frequencies of radiated energy.

HOW RADIATION RADIATES

All radiant energy has three things in common: it has frequency; it has related wavelength; and no medium is needed to transmit the energy. At the low end of the energy spectrum, the radiated energy is purely electromagnetic in character, but at the middle and high ends, it becomes particulate as well as electromagnetic in nature.

Long before people were even aware of electromagnetic radiation of any frequency, we were, of course, using visible radiation as a means of seeing. From the time we thought about the "how" of our being able to see, scientists have argued about light and sight. Early scientists believed that energy flowed from the visible object in one continuous stream like a stream of water from a hose. Sir Isaac Newton proposed that light was only a small part of a broad spectrum, and that light flowed from the visible object to the eye as individual particles in a wave-like motion. He was laughed at for his preposterous hypothesis.

Early in the 1900s, a German physicist, Max Planck, was trying to explain why heated objects gave off a color, and why the color seemed to vary with the temperature. His experiments and conclusions led him to suggest that the dynamic motion of atoms within the material, and the vibration of the related electrons, increased with the amount of input heat energy. He proposed that this dynamic motion caused discrete masses of energy to be shot into the surrounding space with a frequency that depended upon the rate of oscillation of the electrons as they jumped back and forth between electron orbits. Thus, the color—a frequency-dependent visible phenomenon—varies with temperature, and the temperature increases as the input energy is increased.

After much more objective experimentation, Planck proposed his now famous **quantum theory,** which states that heated bodies emit small, discrete masses of energy. Planck called these masses **quanta.** A popular name for them, now, is **photons.**

It wasn't until 1924 that a Frenchman, deBroglie, proved that every moving photon has a definite wavelength that is dependent upon its speed and mass. He further proposed that the speed of all photons is the same, but that the wavelength becomes shorter (higher frequency) as their mass decreases. Therefore, the photons with the greatest mass would have the greatest kinetic energy. These photons with the highest frequency have a special name, **cosmic rays,** because they originate in outer space (the cosmos).

NUCLEAR RADIATION

ATOMIC MASS

While visible, radiant energy is made up of discrete masses of energy called photons, generated in atoms that remain intact, nuclear radiation is made up of the components of disintegrating atoms and an electromagnetic resultant.

In Chapter Two it was shown that according to modern atomic theory, all atoms are made up of protons, electrons, and neutrons. Because the number of protons in every element differs from those of every other element, the number of protons is used as the **atomic number** for that element. The atomic number is the ranking of that element in the order of the elements from the lightest to the heaviest. It might be argued that if the number of electrons is always equal to the number of protons, then the number of electrons is also the atomic number. This is not universally true. When an ion is formed, one or more electrons are gained or lost, but the element remains that element. When the number of protons change, the element changes.

While the number of protons and electrons in the nonionized atoms of a given element are always equal and of the same number, all atoms of a given element do not possess the same mass. For example, carbon has an atomic number of 6, indicating that a carbon atom contains 6 protons (and 6 electrons). However, the atomic mass of carbon has been measured as 12.01115. It might seem that, somehow, the 6 protons plus 6 electrons could equal 12.01115. Such is not the case: it has been shown that electrons contribute practically nothing to the weight of the atom, but that neutrons contribute equally with the protons. But 6 protons and 6 neutrons do not equal 12.01115 either; there must be something else influencing the mass measurement. There are other carbon atoms containing both fewer than 6 neutrons and more than 6 neutrons. Atomic studies indicate that carbon has a total of six different atoms, all with the same number of protons and electrons, but different numbers of neutrons: 4, 5, 6, 7, 8, and 9. Atoms of the same atomic number (6 in the case of carbon) but different atomic mass (because of the neutrons) are called **isotopes.**[1] It is only a coincidence that carbon, with an atomic number of 6, has six isotopes. There is no relationship between the atomic number and the number of isotopes.

Isotopes are identified according to the sum of the protons and neutrons in the core, with the sum printed as a left-hand superscript to the elemental symbol and a right-hand superscript to the full name. The number of protons is written as a left-hand subscript to the elemental symbol. For example, one of the isotopes of nitrogen with the atomic number of 7, which has 6 neutrons in the core, is written $^{13}_{7}N$.

With very few exceptions, all elements have multiple isotopes. The six isotopes of carbon are:

Protons	Neutrons	Isotope
6	4	$^{10}_{6}C$
6	5	$^{11}_{6}C$
6	6	$^{12}_{6}C$
6	7	$^{13}_{6}C$
6	8	$^{14}_{6}C$
6	9	$^{15}_{6}C$

In any collection of atoms of the same element, the distribution of the isotopes will be such that there will be an average mass (for carbon, 12.01115), which is called the **atomic mass** of that element.

An average distribution for a group of carbon atoms is:

Isotope	Percentage
carbon^{10}	0.0001
carbon^{11}	0.0001
carbon$^{12}_1$	98.885
carbon^{13}	1.115
carbon^{14}	0.0001
carbon^{15}	0.0001

With the percentage so heavily weighted at carbon^{12}, it is not hard to see why carbon is considered to have a weight of 12.0 for most inexact calculations.

Translating the above percentages into actual numbers, we find that if we could count a group of 100,000 carbon atoms, there would be 98,885 atoms

---

[1]All isotopes of a given element undergo the same chemical changes, because they have the same number of electrons, which are involved in chemical reactions.

with a mass of 12 and 1115 atoms with a mass of 13. We might or might not find any of the other isotopes because at those percentages, the chances of finding them even in 100,000 is very small. Nevertheless, they do exist.

STABILITY

All too often in the minds of students, the word isotope is synonymous with radioactive. However, all isotopes are *not* radioactive. The lighter elements from hydrogen (atomic number 1) up through bismuth (atomic number 83) are normally stable; that is, nonradioactive. There are, however, some few exceptions found in nature, such as carbon[14]. From polonium (atomic number 84) upward, all isotopes are radioactive to an increasing degree. The latest ones to be discovered are so radioactive, that they do not exist naturally but must be made. They fly apart in a burst of energy almost as soon as they are formed.

The reasons for these bursts of radiologic energy from the core of an isotope are more easily understood when it is recalled that all protons are positive charges, and it is a law of nature that, while unlike charges attract each other, like charges repel each other. Scientists have argued, hypothesized, proposed, and discarded theory upon theory, trying to explain how several protons can exist together in a nucleus of stable, nonradioactive atoms without that nucleus flying apart due to the repulsion of the like charges for each other. No one has really explained it as yet, but the best hypothesis to date is that neutrons must play an important role in keeping that stress-packed nucleus together. There are several observations that suggest this conclusion: no element containing more than one proton exists without neutrons being present; more protons in the nucleus are related to more neutrons, to account for the atomic mass; and where proton to neutron ratios are either very high or very low, radioactivity has been observed from either neutrons or protons exiting from the atom.

It would appear that when the neutrons and protons are in adequate ratios, they keep each other's like-charges apart so that the normal interatomic cohesive forces can keep the particles from being repelled out of the core.

In 1950, a physicist named Mayer proposed that the nucleus was arranged in energy levels similar to the electron orbits on the outside of the atom. He showed that nonradioactive elements or isotopes occur whenever the number of protons or neutrons in the core is equal to 2, 8, 20, 50, 80, or 260. When both the protons and neutrons are equal to one of these numbers, the nucleus is quite stable.

HALF-LIFE

When the proton or neutron content of the nucleus is not one of the above numbers, and the ratio is too much one way or the other, the old, familiar law of balance in nature becomes effective and the nucleus throws off the unbalancers. The result is radioactivity.

The rate at which the core throws off the offending particles is directly related to the amount of unbalance. Greater unbalance speeds the rate of throwing off, causing radioactivity. As the unbalance approaches balance, the rate of throwing off decreases continuously so that the atom is said to approach balance **asymtotically.** This type of approach can best be described by the old riddle, "If you are standing ten feet from a wall, how many times must you step forward in order to get to the wall, if you only go half way each time?" The theoretical answer is, of course, "You will never reach the wall, whether the distance is ten feet, one foot or one hundred miles." The distance has nothing to do with the answer, because each time you go only *half way,* and *half way* gets less each time, so that you never go the last little bit. This is an asymtotic approach to the wall.[2] So, with radioactivity, the longer the decay process goes on, the slower it approaches balance. Because the rate of decay is always changing, scientists measure the decay in terms of half way, or **half-life.** Half-life is defined thus: starting at any time (called time zero) *half* of the radioactive atoms in a group will disintegrate in some given length of time. This length of time is the half-life of that element. Again, half of those remaining at the end of that first half-life will decay by the end of the second half-life.

Holding to our definition that half of the material disintegrates during each half-life, we will theoretically *never* reach zero radioactivity. Practically, however, after ten half-lifes, the radioactivity is no longer measurable.

Some Half-Lifes	
radium	1620 years
uranium^{238}	4,500,000,000 years
proactinium	1.14 minutes
carbon^{14}	5,570 years
polonium	0.0000003 seconds

Obviously, no one has ever measured the half-life of uranium^{238} or radium. As with many other half-lifes, these are statistical extrapolations from short-time measurements. They are reasonably correct nevertheless.

RADIOACTIVE TRANSMUTATION

The most common form of radioactive disintegration is that of two protons and two neutrons (helium^4). With the protons gone, the electrons leave also. Because

---

[2]Practically speaking, after a few half distances you are at the wall, but this is not so theoretically.

both protons and neutrons leave the nucleus, and the electrons leave the outer shells, it is not difficult to see that the atom actually undergoes a complete change. So complete, in fact, when it has decayed enough, it becomes a different element. For example, uranium^{234} can go through fourteen different changes to other elements or isotopes of elements before it arrives at nonradioactive, and therefore stable, lead^{206}. Each change has a different half-life.

ATOMIC RADIATION

In decaying from one element to the next lower in atomic number, the unstable nucleus throws off one of two different kinds of particles. A third form of energy is also given off, not *in* the transmutation process but *because* of it. These forms of radiation are **alpha** (a) particles, **beta** (β) particles, and **gamma** (γ) rays.

Alpha particles are made up of two neutrons and two protons. These alpha masses are relatively large and heavy. Because they contain two protons they are positively charged and, therefore, tend to be attracted to or attract negatively charged electrons. When thrown off by a nucleus, they are sometimes called alpha rays, and contain considerable energy. Because they are big, heavy, and positively charged, they do not travel more than a few inches before being neutralized.

Beta particles are high speed electrons from the nucleus traveling free of internuclear forces. The question might well be asked, "How did the electron get into the nucleus? They are supposed to be in the outer orbits only." Electrons are in the outer orbits, and protons and neutrons are in the core. But a neutron is really a proton and electron locked together, so that their charges balance each other resulting in a neutral particle. More evidence supporting this theory is that neutrons have the same mass as a proton, and that electrons have practically no mass at all.

When the electron is split away from the proton, because of either collision or plain unbalance, it travels at the speed of light, 186,000 miles per second. These electrons are dangerous because of their velocity and their ability to cause ionization of other atoms. They have been known to penetrate over a half inch of wood more than 100 feet (30 metres) from the source.

Gamma rays are a form of radiation that is pure electromagnetic energy given off by a nucleus after it has gone through some form of radioactive decay leaving the nucleus highly energized. Gamma rays have no electrical charge and no weight or mass. They travel with the speed of light, contain considerable energy and have considerable penetrating power.

Gamma rays are the primary source of danger to human health and life. They are able to penetrate almost all materials and can cause severe damage to the human body if they collide with tissue or atoms of a living organ. Only several feet of concrete or several inches of lead will stop these rays.

A way to remember which produces what:

A	B	G
L	Electrons	rAys
Protons	T	M
H	A	M
A		A

MAN-MADE RADIOACTIVITY

In addition to natural radioactivity, which is the decay of naturally occurring yet unstable isotopes, there is another type, **impressed radioactivity**. This type, with identical end results, and radiation, is a product of scientific investigation, wherein, by means of huge, electronic accelerating systems, atomic particles can be brought up to extremely high velocities and then used to bombard normally stable nuclei. The results can be a different element which, if the ratio of protons and neutrons causes an unstable nucleus, begins to decay. Entirely new, and until only a few years ago unimagined, elements have been man-made. They are called the **transuranium elements**. They have been given this name because all have an atomic number higher than uranium (atomic number: 92) which, at one time, was thought to be the last and the heaviest of all elements. All transuranium elements have an extremely short half-life and have never been found in nature.

DETECTORS OF RADIOACTIVITY

Since the human body does not have receivers that can sense alpha, beta, and gamma radiation as we can sense photons of light and fire, we must make use of some other means of determining their presence. We have learned to do this by three basic methods, using instruments especially developed for the purpose. They are photographic and fluorescent or scintillation methods, and an ionization technique.

In the photographic technique, photographic film is used in much the same way it is used to take a picture. In normal photographic usage, particles of light energy cause chemical changes in the sensitive compounds on the surface of the film. When properly developed by reacting the surface compounds with other compounds, these chemical changes appear as a picture. The same changes occur when film is used as a detector of radioactivity. The alpha, beta, or gamma rays cause chemical changes in the surface compounds just as their energy relatives, the photons, did. When developed, the amount of change on the film indicates the amount of radioactive exposure. However, this technique does not indicate the intensity or, if it exists, a rate of change in the intensity. It does indicate the total exposure received up to the time the film was developed. Its simplicity, a piece of film, allows it to be packaged as the smallest and most portable of all detectors of radioactivity.

In the fluorescent or scintillation technique, a counter detects the minute flashes of visible light that are given off when electrons in an element, such as phosphorus, are struck by alpha particles. When the electrons are impacted by the alpha particles, they are driven into a higher energy orbit. As they fall back into their original orbit, the energy is released at a visible frequency. These visible radiations can be converted to audible clicks or pops. They may also be electronically converted to pulses, which can be used to operate counters.

The ionization technique is perhaps the most popular. It is known to most as the Geiger counter. In this system, a gas-filled tube with two electrodes is used as the eye or pickup device. One of the electrodes is a metal sleeve and the second a metal rod that is coaxial with the sleeve and insulated from it at one end. One end, opposite the plastic insulator, is closed off with a material that is transparent to beta and gamma radiation. A schematic diagram of a Geiger counter is shown in Figure 13.1.

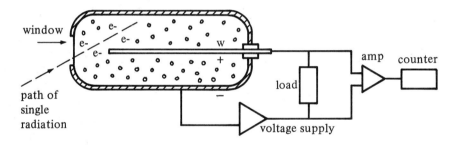

Figure 13.1. Schematic Diagram of a Geiger Counter.

An electrical voltage is impressed onto the cylinder and the coaxial rod so that the enclosed gas, chosen for its ease of ionization, is just below its breakdown voltage. This detector is connected to an electronic amplifier and rate-count meter. When a radioactive particle comes through the transparent end of the cylinder, the gas molecules are pushed beyond their breakdown threshold and are ionized. Being ionized, they conduct electrons across the rod-to-cylinder path and the amplifier and rate-count meter are activated. The output from these components may be a flickering light, a buzzer, a popping sound, or counts on a digital counter.

A form of this counter is also made for alpha particle detection. Because alpha particles are extremely weak, no material has yet been found that will contain the ionizable gas and yet is transparent to the alpha particles. In a unit made for alpha counting, therefore, no end material is used and the gas between the rod and wall is air. Because air has a relatively high ionization potential, this type is far less sensitive than those made for beta radiation.

There is an inherent danger in complete reliance on any instrument used for the detection of radiologic energy. The danger exists because:

• The instrument may not be sensitive to the particular type of radiation, and therefore fail to indicate its presence.
• The instrument, because it is used frequently, may not be operating properly, for example, because of dead batteries.
• Geiger counters can be overloaded and give erroneous readings.
• Improper development of photographic indicators can lead to false indications.

UNITS OF MEASURE

In dealing with the highly complex subject of radioactivity we are dealing with atomic structure and the dynamic effects of its disintegration. The units of measurement of these effects are unfamiliar to most of us, but nevertheless must be mastered. Some of the more common terms used in the field of health physics, dealing with radiation and exposure, are:

Roentgen: One roentgen is equal to the amount of X-ray or gamma radiation that, in 0.001293 grams of dry air, produce ions carrying one electrostatic unit which may be either positive or negative.

Roentgen equivalent physical (rep): This term is equal to one roentgen of about 200,000 volts of X-ray type radiation in soft tissue.

Roentgen equivalent man (rem): This is the amount of radiation of any type that results in the same damage to an adult human as caused by one roentgen of 200,000 volt X-ray type radiation.

Relative biological effectiveness (RBE): This is the biological effectiveness of any ionizing radiation energy that results in specific biological damage relative to the damage resulting from 200,000 volt X-rays of gamma-type radiation.

Curie (Ci): This term is used more often by radiologic chemists and those involved in measuring very small, precise quantities of radioactivity. It is defined as the amount of radioactivity which decays at the same rate as one gram of radium^{226}. By international agreement, this is defined as 22,200,000,000,000 disintegrations every minute. Because this number is so large and almost impossible to handle, except with exponents, a more practical unit has been adopted, the microcurie (μCi). One microcurie is defined as 37,000 disintegrations per second. It is usually written in exponents as 3.7×10^4 dps.

Rad (r): This term is commonly used in connection with biological effects. The rad is defined as the amount of radiation that results in the absorption of 100 ergs by one gram of any material. There is little, if any, difference between a roentgen and a rad.

Million electron volts (MeV): The energy involved in most gamma radiation is in the millions of electron volt range. One electron volt is equivalent to a very small amount of energy, 3.814×10^{-20} cal. One MeV, therefore, is equal to 3.814×10^{-15} cal.

EXPOSURE AND ITS EFFECTS

A safe radiation limit for humans has been determined to be less than 0.1 millirem (0.001 rem) in any seven-day period, or up to 500 millirem per year (whole body dose). An exposure of 0.5 millirem per hour is considered absolute maximum for both animals and humans. A single exposure up to 100 rem will make a human very ill, but for the average healthy adult it will not be fatal. At 500 rem, death is almost certain.

According to present theories, heavy doses of atomic radiation cause the ionization of just about any material they contact. This includes atoms of living tissue where the ionization provides ideal chemical conditions that results in the oxidation of the organic substances called enzymes. These enzymes are the major chemical links needed by living cells to oxidize the fuels required for life. When the enzymes are chemically changed, they cease to do their job, and the cells die. Theory also holds that some forms of atomic radiation attack the cores of the cells directly, killing them through chemical change. Another theory, which adds to the two mentioned above, suggests that high level radiation attacks the bonds of the complex organic chemicals called proteins, and so breaks them down into less complex structures with completely different properties. Without enzymes, proteins, or a proper core, a cell cannot exist.

Once a radioactive isotope exists in the body, having been ingested or formed by the exposure of the body to some form of radiation, that radioisotope is next to impossible to remove from the body. At this time there is no way to eliminate, capture, or in any way make that radiologic cell core nonradioactive. When the half-life of that internal radioisotope is relatively long with respect to the life expectancy of the individual, the hazard can be appreciated. For example, the half-life of strontium[90], reportedly found in cow's milk after fall-out from nuclear testing, is 27 years. If the quantity is high enough or is added to other forms of radioactivity, there naturally is grave danger to human life. To make things worse, radioisotopes tend to collect in specific parts of the body rather than disperse throughout the entire organism. Strontium is a good example again of this type of action. Being a good member of the alkaline family, it always concentrates where another member of that family, calcium, gathers—in the bone and associated tissues. The bone marrow is where the body forms its blood cells. Therefore, when strontium[90] gets into the body it destroys the molecular components of the blood cells. The results may be anemia or leukemia or some other form of cancer.

There are both internal and external effects from radiologic exposure. The external effects, in general, are those of premature aging. There are changes in hair coloring and tissue formation, and the development of eye cataracts. Internally, there are many less obvious effects, but the most devastating one is aging of the blood vessels. In severe cases, external and internal effects will be one or more of the many forms of cancer, such as bone, skin, or thyroid carcinomas, or leukemia. In exceptionally severe cases, such as occurred in the atomic bombings in Japan during World War II, genetic mutations may result.

PROTECTION

Because atomic radiation hazards are generated by and within atoms, these hazards cannot be treated or extinguished like chemically generated radiation, such as fire. There are no extinguishing agents, there is no way to remove the fuel, and there is no simple way to cool it. The radiation continues regardless of any action we may take. We must therefore seek shelter in protective measures and so reduce the exposure to a safe limit. Those protective measures are:

- Reduce the time of exposure.
- Increase the distance from the source.
- Increase the shielding, both around the source and around the area to be protected.

Time

Because atomic radiation is kinetic energy the amount of work it does on any substance can be expressed in simple mathematical terms as:

$$w = \text{time} \times \text{energy}$$

To reduce the work done, we must therefore reduce either the amount of energy impinging on the object or reduce the time during which that energy impinges on the object. A tremendous burst of energy for a very short period of time will do just as much damage as a moderate, but dangerous amount, for a long period of time. We must add "but dangerous amount" because there is an exception inasmuch as we are dealing with a living body. Living cells can and do throw off the destroyed cell cores, the proteins, and enzymes, as long as the number of damaged components is small compared to the number of healthy cells. Just as the whole human body can recover from overexposure to lower frequencies of radiation such as fire or sunburn, so components of that body, such as cells, can recover from long-term exposure to atomic radiation, as long as the intensity is not too great. The problem with atomic radiation is, of course, that we have no built-in receivers that are able to detect it as we do with fire or heat from the sun.

Distance

Where it is not practical to reduce the time of exposure, distance will provide an excellent reduction in the exposure from radioactivity. It will provide a reduction because of the characteristics of alpha, beta, and gamma radiation. Like other forms of radiation, the particles or rays diverge with distance. That is, as the distance increases, the particles or rays spread apart. While this results in a greater area coverage, it also results in less intensity over that area. A good demonstration of this principle is that of a flashlight held about two feet from a wall; the area covered will be relatively small, and the intensity relatively high. When that same flashlight is held four feet from the wall, the area covered will be considerably greater, but the intensity of the illumination on the wall will be considerably less. To physicists this is known as the **inverse square law** and relates to light as well as to radiologic emissions.

This Inverse Square Law is written mathematically as $I = I_s/d^2$, where I is the intensity at the target or body, I_s is the intensity at the source, and d is the distance from the source.

Going back to the flashlight example, if the intensity at the source was 20 lumens[3], at a distance of 2 feet the intensity would be 5 lumens:

$$I = \frac{20}{2^2} = \frac{20}{4} = 5$$

At the four-foot distance, the intensity would be:

$$I = \frac{20}{4^2} = \frac{20}{16} = 1.25$$

While the units would be different for radiation from radioactive material, the ratio would be the same. For example, an intensity of 900 rads at a source would be down to 100 rads at a distance of 3 feet while at 6 feet the intensity would be 25 rads. Doubling the distance reduces the intensity to 25 percent, or one fourth, of the original value.

Shielding

Where neither exposure time reduction nor distance increase is a practical or possible solution, shielding from the particles or rays is the last resort. However, where living tissue must be in the vicinity of atomic radiation, all three methods should be employed.

The type of shielding needed will depend on the type of radiation being evolved. Alpha particles can be stopped by a sheet of paper or even the outer layer of skin. Beta particles can be stopped by a sheet of metal as thin as 1/8 inch (3 mm) of lead, while gamma rays will penetrate 4 inches (10 cm) of lead or up to 2 feet (61 cm) of concrete.

In summary, then, it is clear that there is no portable shielding or type of clothing that can provide adequate shielding from high intensity radiologic emissions. The source must then be enclosed with enough of the proper shielding material.

---

[3] A lumen is a measure of light intensity, defined as the amount of light that a source of one candlepower emits in one solid angle (a ster-radian). Simply put, one lumen is the amount of *green* light emitted by a dissipation rate of 0.00161 watts.

PACKAGING

Due to the rigid controls on radioisotopes, the average person does not come in contact with them except when they are shipped from maker to user. It is in this situation that fire safety personnel can find them in fires.

In general the Department of Transportation of the United States Government requires that a package which has a possible exposure rate exceeding 200 millirads (mr) per hour at any distance or even 10 millirads per hour at three feet from any surface of the container must be shipped by itself on specially marked and equipped vehicles. This type of vehicle is sometimes called a sole use vehicle.

In particular, the Department of Transportation requires special packaging for all radioactive material. The type of packaging is divided into two classes according to the abuse that the package is expected to receive during the transportation process. Type A packaging, for average conditions found in shipping, is expected to maintain the package integrity, protecting the contents from loss or dispersion and keeping its radiation-shielding properties. Type B packaging, for unusual shipping conditions, is designed to protect its contents from severe accident damage in such a way that even with some loss of shielding there will be no escape of radioactivity.

In order to remove the subjectivity from what constitutes a severe accident, the DOT specifies the types and conditions of tests that a package must pass before that design and material can be considered a type B package. These tests include heat exposure for 30 minutes at 1475 °F (802 °C), water immersion for 8 hours, a free fall of 30 feet (9 metres) onto a hard surface such as concrete or macadam, and a puncture test of a 40-inch (102 cm) free drop onto a 6-inch (15 cm) diameter steel pin.

In addition to types A and B packaging, if the material to be transported is judged to be fissionable, the shipper must take adequate precautions to ensure that the material does not attain the threshold of criticality (running away) even under severe accident conditions. The DOT has therefore established three classes for fissile material. "Fissile" is supposedly an easier way of saying "fissionable"; both terms mean that the material is capable of disintegrating by means of radioactivity.

Fissile class I: The only limit to the number of packages of this class that may be transported in any one load is the capacity of the vehicle.

Fissile class II: The limit in this class is the quantity of radiation escaping from the total load. The external radiation level must be lower than the safe minimum, while at the same time there must be assurance that the material will not exceed the threshold of criticality.

Fissile class III: This class must be used when the requirements of either class I or class II cannot be met. This class therefore requires that special arrangements must be made between those who are shipping the material and those who are transporting the material as well as those who are to receive it.

LABELS

In addition to properly designed and accepted packaging, radioactive materials must be identified by a symbol that is quite unique and has international acceptance. This is the purple red three-bladed propeller on a yellow background. It was never intended to resemble a propeller, but is supposed to portray a radiation source (the hub) and the fan-shaped rays of dangerous energy emanating from it. Prudent safety precautions have dictated that the symbol be used wherever radioactive material is present. It is not uncommon, therefore, to see the symbol painted on doors, walls, and containers of all types.

The Department of Transportation requires that one of three labels, each carrying the radiation symbol as the basic unit, be used on all packages containing radioactive material that is being transported in the United States.

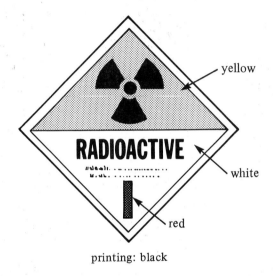

printing: black

Figure 13.2. Radioactive White I.

The first of these, radioactive white I (see Figure 13.2), is a diamond shaped label with a black symbol and printing on a white background. An important feature of this label is that it carries a single, vertical *red* bar printed over the data on the bottom part of the bottom triangle. Radioactive white I may be used only when the dosage rate at the surface of the container does not exceed 0.5 millirem per hour and a zero dosage rate three feet from any point on the surface. In addition, the material must not be fissible class II or class III.

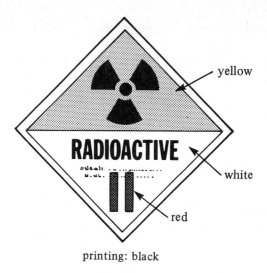

printing: black

Figure 13.3. Radioactive Yellow II.

The radioactive yellow II label is also diamond-shaped with a black symbol, border and printing, but this background is yellow on the upper triangle and white on the lower (see Figure 13.3). Further hazard indication is given on this label by two vertical red bars printed over the data on the lower white-background triangle. Radioactive yellow II must be used when: the radiation at the surface of the container is more than 0.5 millirem per hour but less than 1.0 millirem per hour at three feet from any point on the container; or the material is fissile class II.

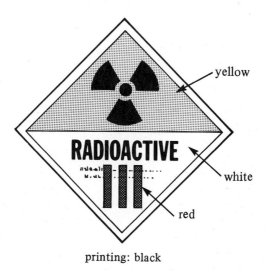

printing: black

Figure 13.4. Radioactive Yellow III.

The radioactive yellow III label is also diamond-shaped with a black radiation symbol and printing and, as with the yellow II label, the background is yellow on the upper triangle and white on the lower (see Figure 13.4). Severe hazard is indicated by three vertical red bars over the data on the lower white-background triangle.

The radioactive yellow III label must be used when the dosage at the surface of the container is more than 50 millirem per hour or exceeds 1 millirem per hour at 3 feet from any point on the container. This label must also be used on any container carrying fissile class III materials no matter what the surface radiation rate is. In addition, when a container of this type is carried in common carriers, the transporting vehicle must carry a warning label on the front, rear, and each side, so as to be clearly visible (see Figure 13.5). The DOT requires that the label be at least six inches on a side.

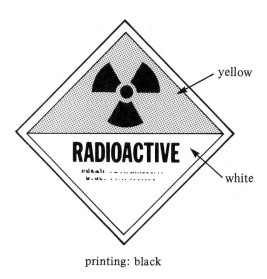

printing: black

Figure 13.5. Label Suggested by the DOT for Vehicles Carrying Containers Marked with the Radioactive Yellow III Label.

FIRE AND RADIOACTIVITY

While it is difficult to remember the exact level of radiation represented by the various labels, a rough conception of levels they represent should be an indication of the relative hazard involved. When these labels are associated with a fire, the Department of Transportation of the United States Government, as well as the shipper, should be notified as soon as possible. In situations where the containers are enveloped by flames but not yet burning, they should be cooled with water as quickly as possible and from as great a distance as the available nozzle pressure will allow. When the containers themselves are burning, again use water from as great a distance as practical but in addition, radiological monitoring should be started and continued as long as the container or containers continue to burn.

In summary then:

1. Evacuate the area.
2. Keep the containers cool.
3. Keep as far away as possible.
4. Keep the radiological monitoring in action.

If there is any doubt whatsoever about what you are doing or what should be done, the Interagency Radiological Assistance Plan (IRAP) should be contacted at once. The mission of this group is to give technical information in radiation emergencies. See the section in Chapter 9 on Special Assistance for details.

OTHER FORMS OF RADIATION

In addition to radiologic energy, there are other forms of radient energy that can be dangerous. Although the degree of hazard is considerably less and the physiological consequences usually, but not always, less severe, these other forms of radient energy can be hazardous. Like so many things, they can be of benefit in small quantities but hazardous in large doses.

- **High power, high frequency radio energy**: High power radio and radar transmitters that concentrate their power by means of special antennas can affect the human body if it is exposed to the electromagnetic radiation at close range.
- **High power, low frequency radio energy**: There is a growing fear even among conservative scientists that emissions from low frequency radio transmitters (in the order of 10 to 100 kHz) can be hazardous if the radiated power is high enough. Some new communication techniques require transmitter outputs in the megawatt (million watt) region. The scientists consider this high enough to be hazardous.
- **Radio frequency and explosives**: Even medium power radio energy at any frequency in the vicinity of long wires used for blasting can produce a hazard. If the blasting wires are of a length that allows them to resonate with the frequency of a nearby radio transmitter, enough voltage can be induced to initiate the blasting fuse and so cause an unwanted explosion.
- **Infrared**: This is the form of energy radiated by heated bodies, and for that reason, often called heat rays. They are not hazardous to the human body except in very large doses such as might be received from overexposure to a heat lamp or from too close proximity to a large flame.

Some people are much more sensitive to infrared radiation than others.

- **Ultraviolet:** Sometimes called ultraviolet rays, this radiant energy is really light that is just above the high frequency end of the visible spectrum. As anyone who has been sunburned knows, the hazard from too much exposure to this form of energy is painful. It can also be lethal. The inherent danger with this form of radiant energy is the delay in the warnings that the exposed tissue produces. A serious exposure can result before any warning is sensed. While ultraviolet energy is always generated by the sun it can also be man-made by sunlamps. The exposure is not less serious because it comes from a man-made device.
- **X-rays:** This is yet a higher frequency than ultraviolet energy. It is closely associated with atomic disintegration but can be man-made as well. There are two types of X-rays, one called "hard" and one called "soft". In man-made devices they are both generated by the bombardment of a metal plate with high voltage electrons. Hard X-rays are produced by the higher voltages and penetrate far more than soft X-rays. Although the soft X-rays are produced by lower voltages, they can nevertheless burn human tissue, and are therefore quite hazardous.

REVIEW QUESTIONS

1. What three things do all forms of radiant energy have in common?
2. Give a simple statement of the quantum theory.
3. What is a photon? What are some of its characteristics?
4. Explain atomic number and atomic mass. How do they differ?
5. What is an isotope?
6. How is radioactivity related to stability? How is radioactivity explained?
7. What is a half-life? Explain how it affects radioactivity.
8. What is radioactive transmutation?
9. Name the three types of radiation from a radioactive atom, explaining each in detail.
10. What are the transuranium elements?
11. How does the popular type of radioactive detector work?
12. Explain roentgen, rep, roentgen equivalent man, RBE, RAD, and MeV.
13. Provide a simple explanation of the effects of radioactive exposure on living tissue.
14. What are the three basic ways of protecting against radioactivity? Explain each.
15. What are the three classes of fissile material? Explain.
16. What are radioactive white I, radioactive yellow II, and radioactive yellow III?
17. Explain in detail the four steps to be taken with a fire involving radioactive materials.
18. What other forms of radiation, besides radioactivity, are hazardous? Why?

Index